New Directions in Packaging

Contributors

Marshall G. Baldwin
Joseph M. Callewaert
Kenneth E. Clark
David J. Freiman
A. W. Koehlinger
Charles A. Lewis
Edwin A. Locke, Jr.
Eric B. Outwater
Dolph W. Von Arx

New Directions in Packaging

American Management Association, Inc.

This book has been distributed without charge to AMA members enrolled in the Packaging Division.

International standard book number: 0-8144-2142-3
Library of Congress catalog card number: 77-131550

Contents

Packaging in Today's Economy

Charles A. Lewis

A CONTINUAL barrage of economic discussions concerning current conditions and short-range forecasts hits the daily newspapers and business and trade journals from authorities in and out of government. I certainly do not want to add fuel to the fires but rather to relate the overall economy of the country to the area of packaging. In so doing, I shall look briefly at where the economy went during the sixties, with major emphasis on what happened in 1969. I shall also discuss briefly where we are today and look at some major issues that could affect packaging in the near future.

We all know that the sixties were called a soaring, sizzling, swinging, scrambling, scaling, surging decade. But no matter

CHARLES A. LEWIS is director of the Office of Consumer Goods, Business and Defense Services Administration, U.S. Department of Commerce.

what phrases or catchwords were used to describe the economic activities of the sixties, a hard look at the facts reveals that the United States experienced an economic expansion that was unprecedented, registering an 85 percent gain in the gross national product (value of goods and services). During the same period, the value of packaging moved from $11.5 billion to $19.5 billion—a gain of 70 percent in terms of current dollars.

These data indicate that there was a more rapid growth in the overall level of economic activity than in packaging. Keep in mind, however, that this comparison between growth in the economy and growth in packaging is in terms of current or inflated dollars. If we take a closer look at these data, we shall arrive at a different conclusion.

If we deflate the GNP dollars to find the real growth of the economy (1960 = $487.7 billion; 1969 = $727.5 billion), we find that it was only 52 percent during the sixties. We also find that price inflation accounted for 33 percent of the 85 percent dollar gain in the GNP as previously mentioned.

Let us look at some price increases in packaging. From the figures available in the Bureau of Labor Statistics (BLS) Wholesale Price Index, we find that metal can prices increased 20 percent during the 1960–1969 period. With price increases of 7.5 percent in aluminum foil, 15.8 percent in metal drums, 17.2 percent in glass containers, and 5.2 percent in corrugated shipping containers, inflation accounted for only about 12 percent of the total dollar growth during the sixties. If we deflate the packaging value by 12 percent, we arrive at a 58 percent real growth in packaging as compared with a 52 percent real gain in overall economic growth.

If we can draw any conclusion from these sets of data, it must be that technological developments and increases in packaging productivity gave container users more mileage for the dollar spent. An additional reason for container "bargains" may be the intensified competition that occurred within and between various sectors of the packaging industries servicing the same markets. We also must recognize the stepped-up trend toward vertical integration and horizontal diversification that has occurred in the sixties within packaging and related fields. In addition, packaging material prices have not exceeded the average price index of all materials, and in many instances have been below the average. The economy of the sixties can be described as a prosperous, expanding giant with new records set almost every year, and packaging experienced the same growth pattern.

We must review the American economy in 1969 in more detail, because developments in the last half of the year unquestionably set the economic climate for the early seventies. In 1969 the economy, as measured by GNP, recorded its ninth consecutive year of expanding activity. The year began with production, income, and prices increasing rapidly. In the course of the year, the pace of activity slowed progressively, although total GNP in current dollars was 7.75 percent over the 1968 level. This increase, however, reflected a 4.75 percent inflationary factor that reduced the real economic growth to 3 percent. In comparison, the 1968 GNP was 9 percent over 1967, with 4 percent accounted for by price increases. These directions indicate that prices were increasing while real growth was diminishing.

In packaging, 1969 shipments valued at $19.5 billion exceeded the 1968 total by 6 percent, and, like the national economy, scored a ninth consecutive year of record-volume business. A cursory examination of some of the major elements of GNP may give some answers as to why the economic tempo slowed perceptibly during the last half of 1969. Although personal income increased about 9 percent in 1969 over 1968, consumer spending grew cautious as the result of uncertainty about the future. Reduced consumer expenditure was largely felt in the durable goods areas, especially in automobiles, furniture, and household equipment.

Packaging did not reflect this downturn in consumer spending, because about 87 percent of all packaging material goes into nondurable goods industries such as food and beverage, tobacco, textile, paper, and the like. The food and beverage industries alone consume almost 60 percent of all containers and packaging materials produced, and consequently packaging is tied very closely to the housewife's food dollar. In fact, it has been said that out of every $20 that a housewife spends for food, $1 to $1.50 goes for packaging.

Total business expenditure for new plants and equipment in 1969 was up 11 percent over 1968, due in part to expectations of further inflation and fears that the federal government would eliminate the 7 percent tax investment credit. (The tax credit was repealed as part of the Tax Reform Act of 1969.)

During the sixties, packaging industries invested more than $4 billion in new plants and equipment to satisfy anticipated user demands and introduce new containers and packaging materials. Although I do not have any specific 1969 profit data on packaging, I have no doubt that the financial picture was, in general, the same as that experienced by all industries.

On the whole, the labor market continued to be tight in 1969, although a significant easing up became apparent late in the year. Unemployment of skilled workers was slightly over 2 percent of the total workforce; the total unemployment rate was 3.7 percent. The packaging segment continued to feel a shortage of highly skilled workers and, in some instances, unskilled laborers as well; the turnover rate among unskilled workers continued at a high rate.

As already stated, inflation accounted for 4.75 percent of the dollar gain in 1969 over 1968, the sharpest annual increase since 1951. A rising labor cost, up 7 percent, was among the underlying causes of the sharply rising prices in 1969. (Labor cost is the single most important component of prices, accounting for 62 to 65 percent of the total.) Strong demand and rising costs pushed consumer prices up more than 5 percent; industrial prices rose nearly 4 percent. The economic slowdown that occurred in the last half of 1969 resulted in part from a planned reduction in federal spending, mainly for defense.

In addition, fiscal and monetary restraints imposed directly and indirectly by the government were also intended to slow down consumer demand. As mentioned previously, the effectiveness of these combined restraints was reflected in a reduced demand for housing and durable goods. The result was that the real growth in demand for output in the fourth quarter of 1969 was at zero level as compared with the third quarter. This reduced demand extended into the second quarter of 1970, but at a lower rate.

Packaging activities in the second half of 1969 were normally higher than in the first half, and were also above the comparable period in 1968. Business in the fourth quarter was down a seasonal 2.5 percent from the third quarter. Preliminary and fragmented information for the first quarter of 1970 indicates that packaging business will be lower than the final quarter of 1969. This is normal.

If we accept the statement that packaging has a tremendous impact on our economic way of life, then, conversely, we must accept the statement that the economic climate has a tremendous impact on packaging. We know that packaging growth is closely correlated to the expenditure of the consumer dollar for food and related products, and that employment, salaries, and profits in packaging closely follow the composition and direction of U.S. totals. New or expanding demands in the marketplace result in expenditures for new capacity, new plants, and new equipment, increased competition within and between packaging industries, and increases in R&D spending for new containers and packages.

We also know that more new products are created in a favorable economic climate and that new industries arise to fulfill new consumer demands. Furthermore, the packaging economy will reflect the demand of over 300,000 manufacturing establishments and over 2 million retail and wholesale establishments, whose managers recognize the tremendous impact that packaging has on the marketability of their products. Of course, we cannot forget the U.S. government, the largest single user of containers and packages. It is difficult to determine exactly how much money was spent on packaging directly or indirectly by the federal government, but it is estimated that between 5 and 10 percent of the product purchase price can be applied as packaging costs.

The economic prosperity of the sixties made packaging a major consumer of basic industrial products such as steel, paper and paperboard, wood, plastics, aluminum, textiles, and glass. This favorable economic climate also resulted in a net gain of about 5 percent in the total number of plants manufacturing containers and packaging materials. More than $4 billion was spent by packaging manufacturers during the 1960s for new plants and equipment to service expanding markets that especially reflected consumer demands. In addition, total employment in the packaging field expanded by about 15 percent during the sixties; wages and salaries increased almost 50 percent; the cost of raw materials and supplies increased about 40 percent; and the ratio of cost of labor and materials to value of shipments in 1969 remained approximately the same as in 1960.

I am not sure whether today's packaging market was created from "demand pull" or from "packaging push," but the resulting market has created one of the largest and most dynamic industries in the United States, ranking closely behind such economic giants as construction and the food, textile, chemical, and automotive industries. Not all segments of packaging have experienced, or shared equally in, total economic growth, however. For example, wood and textile containers, in general, have grown at a below-average rate, whereas plastics, composite, and foil containers and packaging materials have shown an above-average rate of growth during the sixties. The big question now is, "Where is the economy going in 1970 and how will it affect packaging?"

As far as the direction of the economy is concerned, I believe that a paragraph in *The Wall Street Journal*[1] shows the complexity of the question. It said, "It all depends on

[1] Alfred L. Malabre, Jr., "How's Business," *The Wall Street Journal*, March 20, 1970.

which economist is talking—a serious recession looms; a mild recession looms; no recession looms but business will grow very slowly; business will boom." It further states, "If nothing else, the diversity of opinion helps underscore the fact that nobody, however expert, really *knows* what impends on the business front." Further divergence of opinion can be seen in statements by those who point to a continuation of the slowdown in residential construction, monetary restraints, and cutbacks in automotive and appliance output. This group further points to the fact that the downward slide in industrial production reached its seventh consecutive month in February 1970, the longest such decline since the 1960–1961 recession. But on the positive side, the Council of Economic Advisers to the President states that the 1970 economic prospect is for a current dollar gain of 5.5 percent over the 1969 volume, with GNP achieving a $980–990 billion range. Support of this view can be found among many businessmen, economists, and consultants.

In general, industrialists are more optimistic about business prospects than are the economists or forecasters. For example, an anticipatory survey conducted jointly by the Department of Commerce and the Securities and Exchange Commission reveals that businessmen expect to spend 10.5 percent more for new plants and equipment in 1970 than they did in 1969. Durable goods manufacturers predicted a sales gain of 7 percent in 1970 over 1969, and nondurable goods companies anticipated a 6 percent increase. Most businessmen indicated that they expected no change in the first six months of 1970 business, but saw a step-up in activity during the last six months. One leading industrialist in the packaging business stated, "All indications are that the packaging industry, which has the advantage of primary dependence on consumer buying of nondurable goods, will grow during 1970 at a considerably faster rate than the GNP." In general, businessmen expect costs, wages, and prices to be at a higher level in 1970 than in 1969, but to increase at a slower rate.

My view on the level of packaging in 1970 is expressed in

the Department of Commerce publication, *U.S. Industrial Outlook—1970.* The prediction for the value of shipments for packaging in 1970 was $20.5 billion, a 5 percent gain over 1969. Early and very preliminary indications are that the first-quarter 1970 packaging business will approximate the 1969 physical volume. However, if allowances are made for price increases, the packaging business is now slightly behind the 5 percent prediction for 1970. So the prediction of a 5 percent gain may possibly be a shade high but certainly within the realm of possibility, *if* the economic upturn viewed for the last half of 1970 occurs. In any event, the heavy dependence of packaging on the relatively inelastic demand for food and other nondurable goods insures a growth in packaging that could exceed the real gain in total economic activity. We can all, then, be optimistic about the short-term and long-term growth prospects for packaging.

The Marketing-Packaging Interface

Dolph W. Von Arx

Marketing, with its "umbrella" of functions, has traditionally been considered to embrace the areas of advertisting, sales, merchandising, and distribution. Marketing has developed and matured in the last two decades, however, and some areas that were at one time only distantly related to marketing have now become interrelated with it.

Packaging has also developed and matured, and has become an increasingly vital element in the total marketing picture. It is only recently that the packaging function has gained the stature of a fully organized and functional department in many companies. Packaging has, for the most part,

DOLPH W. VON ARX is vice-president and director of operations at the Checkerboard Food Division of the Ralston-Purina Company in St. Louis, Missouri.

evolved from the purely mechanical function of enclosing an end-product to an activity with a marketing significance that must be recognized as today's industry looks to the future. In essence, packaging has become a business activity that marketing cannot ignore but must rather embrace in order to perform an effective job.

The interface between marketing and packaging groups can make or break a product line. Let us look at the brewing industry, for example. A major effort was placed behind the "tapper" package by a well-known brewery. After several reversals, the product was withdrawn from most markets. On the other hand, several breweries, with the help of packaging suppliers, introduced the easy-opening pull-tab cans a number of years ago. This innovation met with widespread success, and the industry has accepted this as a standard package. These were packaging decisions and, we might add, marketing decisions; the interrelationship is clear.

Packaging actually encompasses many activities within itself, including machinery and mechanical systems, engineering, research and development, specifications, quality control or quality monitoring, structural design, graphic design, design research, copy, legal, and purchasing. All these activities fit together and comprise the packaging team. In the same manner, packaging is an important part of the marketing team and has an important position in the successful marketing effort.

Things change, and, as has been the case in many other phases of business activity, the changes that have come about in most areas of packaging have been occurring at a fairly rapid pace. We can anticipate that future changes will occur at an even more accelerated pace. It is important to anticipate these packaging changes and improvements in our marketing efforts in order to keep our customers buying our products, to attract new customers, and to enable us to continue to market new products. In the field of consumer products, Ralston Purina has stated its position as being marketing-oriented—packaging will play a major role in this effort.

Communications: The Problem

There is no question that improved communications between the packaging and marketing groups can produce better results, and, we might add, better year-end profits. It would be naive to say that communications between packaging people and marketing people are crystal clear; unfortunately, packaging is often positioned as an orphan in the total marketing mix. Marketing people, particularly young product and brand managers, often become totally involved with the advertising and sales promotion aspects of the marketing program. Marketing men who have only a limited knowledge of the true function of packaging often look upon these advertising functions, which include the creation and execution of television commercials and print advertisements, as *the* vital elements of the marketing mix.

Perhaps packaging people (both suppliers and members of corporate packaging staffs) should spend more time selling the value of their function and end-product. It is important to communicate to marketing people the true importance of packaging essentials. After all, a package should be an extension to the point of sale of the product-line advertising. In essence, a package is an ad on the shelf, and this is where the final decision to buy is taking place.

What can be done to improve communications between marketing and packaging people, so that more effective use will be made of the packaging function? The answer is not a simple one; however, there are two very basic steps that can be taken: The first step is to establish packaging objectives; the second step is to organize to meet these objectives.

Establishing Objectives

First, we shall define objectives for each packaging effort. These objectives should be spelled out for each established product line as well as for new products and product concepts that are still in the development stage. After these objectives are clearly defined, they must be communicated to the appropriate packaging and marketing groups so that plans can be made to achieve them. In order to establish meaningful objectives, it is important to recognize the basic requirements of a convenience product package; these basics are too often overlooked in the development of packaging plans. For a moment, let us review the requirements that a package must meet:

1. It must *contain* a product.
2. It must *carry* a product through the distribution system.
3. It must *protect* a product.
4. It must *keep* a product fresh and usable.
5. It must provide a *storage* (or *utility*) function.
6. It must *merchandise* and *sell* a product.
7. It must *resell* a product.

We shall not dwell on the fundamental requirements of *containing* and *carrying* the product. Admittedly these are very important, because they are the most basic elements of packaging; however, the packaging industry can, for the most part, provide the proper basic package. It is when we move into the areas of *protecting* and *keeping* the product that greater sophistication is required. The marketing people must provide the packaging people with information about the distribution system through which the product and package will move to the consumer. The package must protect the product so that it is literally intact and undamaged when it reaches the consumer. To keep a product fresh and usable is a difficult task, because this function deals with the ability of the package to retain, to the greatest degree possible, the original state of the product at the time of its manufacture. Moisture changes, oxidation, and time all play a major role in the keeping requirements of a package. The question, "How much shelf life must my package have?" is one that must be answered by the marketing group. The packaging department will then determine the kind of package that is needed to meet the requirement.

If the product is of a kind that is used gradually over a period of time, then the *storage* function enters the picture. Steps must be taken to maintain the product in the best state possible. Today there is an unquestionable concern with and desire for greater convenience in goods and services. The marketing-packaging relationship tightens still further as the need to offer convenience becomes almost a necessity. The need for a package also to perform a *utility* function is an increasingly popular concept, as evidenced by the numerous products available today that are packaged in heat-in bags, or in heat-in and cook-in pans.

The *merchandising* function of the package should not be understated; anyone involved in the food industry today is familiar with the literal struggle that exists in this phase of marketing. How does the package fit into this struggle? It can work at being a successful merchandiser. Merchandising is the segment of the total marketing function that provides the most intimate link between the product and the purchaser of the product—it is the package and product displaying itself, literally saying, "Pick me up and take me with you." It is the *extension* of advertising to point of sale.

The package's ability to be a good merchandiser depends on its size, its shape, its appearance, its color, and its character. But there is more, because the package has to be a salesman, too—it has to *sell.* Advertising, regardless of whether we conceive of it in terms of packaging or in terms of a television commercial, sells the consumer. There are many elements that work together to make the package a good salesman, and they are not put together as easily as it may seem to the average person.

We cannot stop at selling, however; the package has still another job—it must *resell.* To resell a product, one must maintain a high level of consumer satisfaction. It is necessary, therefore, constantly to upgrade our product packaging, from a utilitarian as well as from an esthetic standpoint. To maintain packaging excellence, we must rely on advance planning and packaging research.

Most manufacturing companies are not engaged in packaging research. In most cases, budgets do not permit this type of activity unless the company cannot find solutions to a given packaging problem by other means. The vast majority of packaging research is carried out by companies that are engaged in manufacturing and supplying packaging equipment and packaging materials.

Packaging development is another story. New or revised packages are developed by outside consulting firms, development departments within a company, or packaging suppliers. The choice of which avenue to use in the development of a package depends on the circumstances involved. In some cases, all three avenues may be pursued. For the most part, however, the development is usually accomplished by the supplier, with the supervision and guidance of the user.

Underlying these basic requirements are the very important considerations of cost and availability. We must maintain a proper perspective of the requirements relative to the total profitability of the product, which is, after all, our ultimate objective. Adequately protecting a product at the lowest possible cost is at the root of all packaging development work.

Organizing to Meet Objectives

This leads us to the second basic step in insuring packaging excellence—organizing to meet our objectives. Well-thought-out objectives and plans too often fall prey to poor organization within a corporate structure. The responsibility and accountability of the various line and staff groups must be clearly delineated in terms of how they relate to the key functions of the marketing mix. Marketing people set sound packaging objectives and then too often find themselves, either by choice or by chance, completely divorced from the activities and plans that are formulated to meet those objectives.

It will be helpful here to illustrate two fairly common

Exhibit 1
The Insular Approach

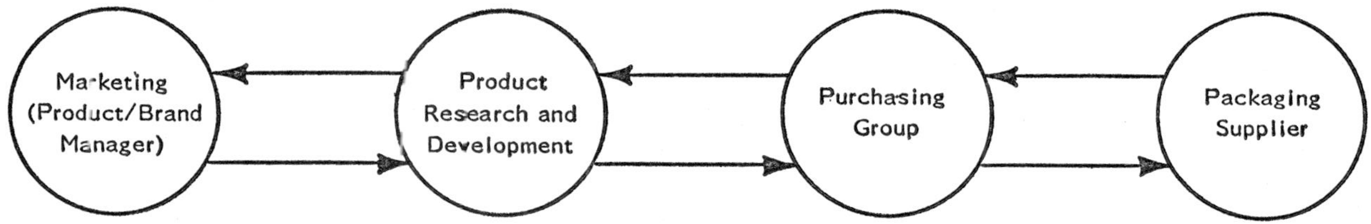

Exhibit 2
The Superinsular Approach

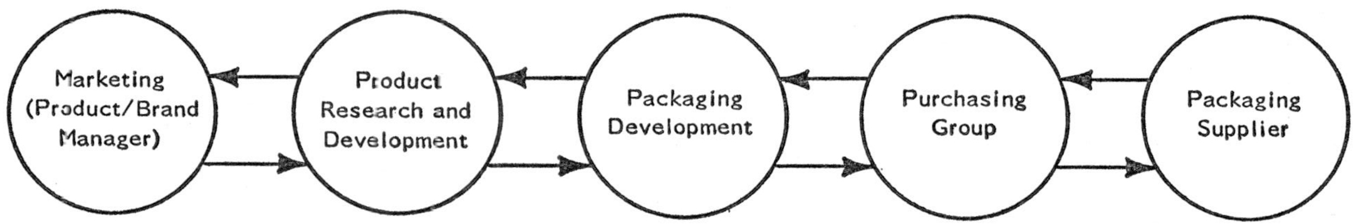

approaches to organization structure in packaged-goods companies. The "insular approach" is illustrated in Exhibit 1. This organization structure insures that the marketing group is completely insulated from the packaging supplier. The layers of insulation are the product research and development group and the purchasing group. Marketing communications are channeled through these two layers to the supplier.

The second approach, which we shall call the "superinsular approach," is illustrated in Exhibit 2. Here we have cleverly added still another buffer to the communications process between marketing and the packaging supplier. This buffer, packaging development, further divorces the person concerned with objectives (the product/brand manager) from the group that executes the packaging plan (the packaging supplier).

These two approaches to organization insure one thing—there will be a communications gap. Establishing a clear set of objectives is not in itself an answer to the problem of communication; we must provide a realistic organizational framework within which these objectives can be met.

We are not questioning the value of the product research and development, purchasing, and packaging development groups to the development of efficient and effective packaging. However, we are questioning the manner in which they are organized to pursue the development plans. It is increasingly common to question established organizations in these times when we see local, state, and national governments being challenged by various dissident groups. A responsible part of questioning an established organization or procedure, however, is providing alternative solutions to problems that precipitate the concerns and questions.

Communications: A Solution to the Problem

Therefore, we shall offer an alternative, if simplistic, solution to this communications gap that has developed between marketing and packaging suppliers—we shall call this proposed solution the "open door approach to packaging excellence." As illustrated in Exhibit 3, this organization structure sets up marketing as the hub of a communications wheel. The wheel arrangement involves each staff group that is important to the development of the objectives and the execution of the final plans. The product or brand manager has direct communication with the packaging supplier as well as with other important staff groups.

Exhibit 3
The Open Door Approach

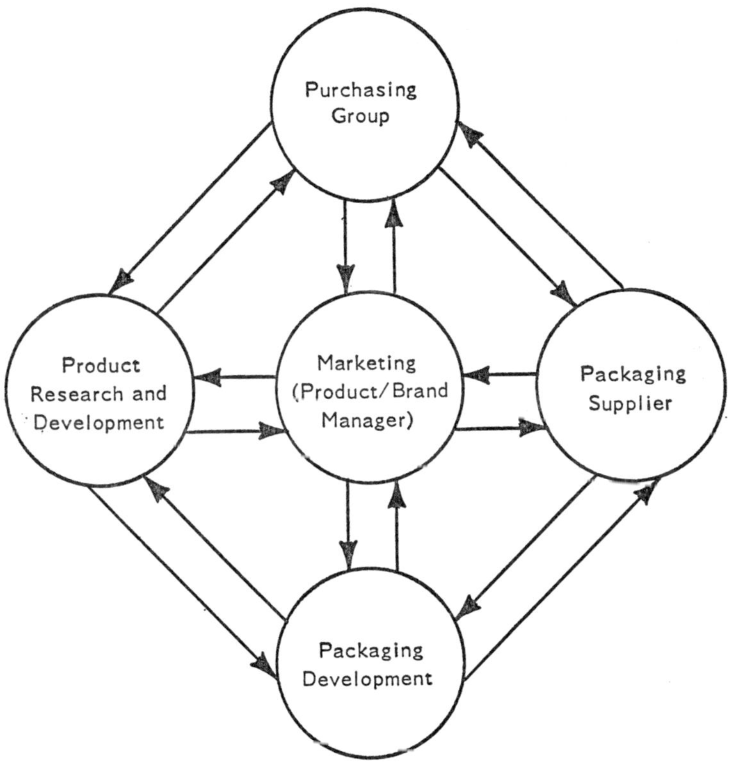

It is important to note that the open door approach can succeed only if the responsibilities and the accountabilities involved are set down clearly for each staff group. This is necessary in order to avoid duplication of effort or failure to execute a specific step in the planning process. The open door approach to packaging forces the product or brand manager to become more directly involved in the packaging development process, and this involvement considerably enhances the value of packaging as part of the total marketing mix.

The Role of the Packaging Supplier

We have talked about the establishment of objectives, as well as about organizing to meet these objectives. It will be productive to examine the role of the packaging supplier as perceived by many marketing-oriented packaged-goods companies. We must keep uppermost in our minds the fact that the primary function of the packaging supplier is to execute the packaging strategy and plan, which flow from the basic packaging objectives for any given product. The speed with which these functions are executed depends, for the most part, on the nature of the product, the production requirements, and the objectives set forth for the product.

It is important that we stress this point—when a product reaches the place where the package must be developed and eventually specified, it is imperative that the requirements be clearly stated. If the requirements or the product's objectives change, these changes must immediately be communicated to the package supplier so that he can make necessary adjustments without a loss of time and effort.

The materials used in packaging are usually developed by the manufacturer of these materials. This is not to say, however, that the incentives for meeting various material requirements will be provided exclusively by the suppliers of these materials. Many of the ideas and directions must be provided by the user to meet certain requirements of his product, just as many of our future products will be determined by the needs and wants of our potential customers. For example, the paperboard material used to make our dry pet food cartons (Cat Chow, Cat Dinners, Dog Chow, Puppy Chow, and Chuck Wagon) was an outgrowth of a presentation by our purchasing group to suppliers. This presentation illustrated the need for a linerless carton that would not stain and, hence, would maintain an attractive appearance on the shelf.

Our suppliers went to work on the project and developed a paperboard that provided the characteristics we needed. It was a breakthrough in pet food packaging and it gave us a marketing advantage. By eliminating the waxed glassine liner, we not only saved the direct cost of the liner material; we also eliminated having money tied up in the inventory and handling costs of waxed glassine. Another very important advantage was that we were able to package the products on high-speed equipment (which cannot be done with cartons that require a liner), thereby reducing the unit cost for packaging. A final marketing advantage was that we were able to provide consumers with a package that was more convenient, because they no longer had to contend with a liner.

Suppliers can also offer assistance in the areas of market research, design research, and the creation and testing of design and structure. These are the building blocks that help to form a good merchandising package. Once again, it is worthwhile to point out that the primary objective of the packaging supplier is to create sales opportunities for his company. In this regard, an overly optimistic view can result in the misdirection of efforts, so it is necessary to analyze carefully the value of the assistance they offer. With this in mind, it is important for us to be sure that in our acceptance of our packaging supplier's assistance, we have not overlooked the competitive services offered by other companies. There is always a danger of allowing ourselves to become stereotyped by a given approach.

The preceding statements were not intended to create an

undue feeling of apprehension; for the most part, these services are designed to assist in the successful marketing of a given product. The efforts of packaging companies are, in most cases, very professional and often more reliable than assistance that is obtained outside the packaging field. In other words, the suppliers of packaging materials want success as much as do the people who are marketing the product. The results are obvious—additional sales for both companies.

Graphics and Packaging

As an illustration of the dependence of packaging on marketing, we find that many terms that are part of the vocabulary of marketing people and packaging people today are terms that did not even exist several years ago. Such terms as "tachistoscope," "visiometric comparator," "ocular camera," "angle meter," and "blur meter" are recognized by these people as the names of instruments used in the perceptual evaluation of graphics and structure.

There is no doubt that an already complex interlacing of marketing and packaging will become even more complex in the future. It is in the area of package design and design research that a greater amount of cooperation may first be achieved, because the need for facts and information regarding design must be fulfilled without the delays that we have been able to afford in the past. With new products being conceived and developed at an ever increasing rate, the need to act swiftly and decisively becomes more apparent.

Package designs, like package materials, are created by various sources. Usually, however, they are agency-created, supplier-created, or obtained on the outside from a freelance or other professional designer. There is a concentrated effort being made by people in the packaging field toward developing graphics that will sell. In order to come up with design creations that can sell effectively, it is necessary, at the very beginning, to state the objectives, the purpose involved, and the limitations that must be adhered to.

By moving to a specific area related to graphics, let us examine a procedure that has worked well for our company in regard to the litho positives that are used to produce graphics in their printed form. It has long been a policy of packaging suppliers to hold title to, or actually to own, the positives that they have produced (or have had produced), even though the service and the materials were paid for by their customers. A number of years ago, we adopted a policy under which we have our own positives produced, and then supply duplicates to the packaging suppliers who use them. In this manner, we not only have complete control of the production of the positives, we also own the material involved and therefore can provide duplicates at a low cost and with great speed when necessary. This policy—which, by the way, has been adopted by other major companies—has proved itself by giving us flexibility and by meeting our needs in quantity reproduction at optimum cost.

We in packaging and marketing are sometimes so involved in the everyday details of our jobs, that we inadvertantly fail to look to the future. It is important to stop and reflect, and also to *project,* because the very fact that we are in a dynamic environment requires us to think in terms of comprehensive advance planning. It has been said that "packaging is a catalyst in converting natural wealth into distributable form." This statement can be made more inclusive by appending the conclusion: "And when married to the marketing approach, it brings about the successful cycle of business activity." We who are engaged in packaging and in marketing deal with materials, with factors, and with forces that create the kind of interest that stimulates improvement and change. Things do change; things are changing; we can and must be ready to react to and move rapidly with change and the changing needs of our customers.

The Human Element in the Packaging System Complex

A. W. Koehlinger

For our purposes, we shall define a packaging system as a packaging operation that involves the use of two or more types of packaging materials and two or more types of packaging machines. If any one of these components is inadequate—be it the packaging material or the machine—it will affect not only the operation of the remaining components of this system but the production line as a whole.

It follows from this definition that a rewarding systems venture can be achieved only if the components, the packages and the machines, are successfully melded and complement one another. We shall not dwell on the consequences of the use of machines not designed for packages (as a packaging

A. W. KOEHLINGER is director of marketing for R. A. Jones & Co., Inc., in Cincinnati, Ohio.

material supplier might describe it) or packages not designed for the machines (as a machinery supplier might describe it). The consequences, regardless of whether it is the package or the machines that are at fault, are inefficiency, expense, and a terrible headache (as the customer would describe it—in one of his calmer moments).

Machines and packages in themselves are just that—machines and packages. The theme of the following remarks, which will be primarily descriptive rather than prescriptive, can best be stated as: "*People* make machines and packages into a packaging system." What follows, then, is a description of potential problem areas that are caused by people and can in turn be solved only by people.

Who are the people? Basically there are three broad groups, the first of which is the many-peopled, organization-charted, job-descriptioned customer. The second group consists of the packaging material suppliers, who provide such items as bottles, cans, cartons, cases, films, and labels. The third group consists of the "hardware" suppliers, who provide the wide variety of packaging machines, as well as conveyors, diverters, convergers, accumulators, and code daters. The problem, then, is how to coordinate the activities of the people who work in these different groups. After all, the firms and the people who work for them are geographically separate and differ in such respects as corporate goals, methods of compensation, and background. To succeed in eliciting genuine cooperation from such diverse groups is not a simple matter.

Let's take a look at the basic motivations of these groups and how these motivations affect the packaging system. The customer wants a successful, reliable, on-stream, on-time packaging system. The packaging material suppliers and the hardware suppliers want the same results. There is obviously more to motivation than the end toward which it is directed. Each supplier is attuned to serving the customer; he recognizes that his long-term success or failure rests upon his ability to fulfill customer needs. In the normal, day-to-day customer-supplier relationship in which he is the only supplier involved, the supplier can readily measure his performance in fulfilling the needs of his customer.

Packaging systems building brings a new twist—complementary, cooperative efforts among the different suppliers become mandatory for the life of the project. Suppliers are not generally accustomed to or motivated toward cooperative efforts with other suppliers; other suppliers are thought of either as competition or as potential competition. This observation is so obvious that I hesitate making it, but there is no doubt that unless the customer fully understands this supplier inhibition at the start of a packaging systems venture, the entire project will probably be not only tedious but also less than completely successful.

The customer must not assume that all he has to do to rally his suppliers is to state in a sketchy manner that they should work together. Suppliers with no more than vague instructions to cooperate from their mutual customer will tend to do so at least at arm's length. Each supplier must understand how his own cooperative efforts with the other suppliers will ultimately benefit his customer, and how his customer will recognize, measure, and reward him accordingly. The ground rules of customer-suppliers and supplier-supplier relationships must be fully understood by all participants at the outset.

On occasion, this is easier said than done, not because the suppliers are uncooperative or not ready and willing to be instructed, but because it has not been determined who the customer is. This is our second potential problem area—corporate impersonality. When I am involved in the drafting of legal contracts, I invariably refer to a corporation as a "he," and the attorneys invariably substitute "it" for "he." That is precisely the kind of problem I am referring to: Sometimes the customer appears as an "it" to the suppliers. "Its" don't make decisions; they don't coordinate; they don't arbitrate; and they don't motivate.

Suppliers do not serve "its" or XYZ Corporations; they serve the people who are employed at XYZ Corporation, particularly those people who measure their performance and provide them with their rewards. If all of the suppliers who are involved in the total packaging system are not receiving their instructions from the same, single place on the organization chart, there is a good chance that an "it" situation exists. For the best results in obtaining an efficient packaging system, there should be one person, whether he be the project manager, the venture manager, or the systems engineer, who is the customer. *He* must be the single door of entry to all of the suppliers involved.

A third potential problem area evolves from a combination of the two that we have discussed. Many packaging systems, because of the unique and pioneering aspects involved in creating them, cannot be fully specified, described, and pinned down at the outset. The interrelationship between the various components (materials and hardware) cannot be fully recognized until the system starts taking shape. Perhaps this is the reason that the objectives are often pushed aside or forgotten at the start. There is a distinct difference between "how" the system evolves, "how" the system is made to work, and "what" the system must do. Because the "how" is necessarily quite nebulous at the start, we tend to forget to establish the "what." We should establish "what" the rules of the game are before game time. If necessary, "how" the game is played can develop while playing.

What should the ground rules of the customer-supplier relationship be? An obvious generalization is that the parameters of responsibility should be defined as precisely as possible. The more nearly unique the packaging systems project is, however, the more difficult it is to set these parameters. Thus the more nearly unique—the more pioneering—the packaging systems project, the shorter the "arm" must be in the "arm's length" relationship between customer and suppliers. In this situation, the importance of mutual trust, of establishing a sound mutuality of interest at the outset is even greater. New consumer products are being introduced at a rapid rate today; it seems that the day of the comfortable, leisurely pace is past. Mutual trust and interest are particularly relevant in these conditions of rapid change, because the capabilities of the packaging system may have to be altered before it is even completed.

Perhaps because of some of the pitfalls involved in the evolution of a packaging system, there seems to be a trend toward partial or complete transfer of the systems responsibility by the customer to one of his suppliers. There are understandable differences of opinion as to what this accomplishes. It can relieve the customer of some of his overall coordinating efforts, and it can prompt him to be particularly definitive about what is to be accomplished at the outset. In certain situations, it may allow the customer to reduce his financial risks in the event that a successful system is not created. If the customer is not staffed for systems creation, if one of the suppliers has had more experience in setting up this kind of system than the customer, or if there are few unknown or unproven concepts involved, a supplier-managed systems development may be appropriate.

In some instances, only a partial transfer of systems responsibility is made by appointing one of the suppliers to a coordinating or a consulting engineer role. His out-of-pocket cost will generally be less if he follows this course than it will be if he transfers the financial risk for the complete system to one of his suppliers. This is understandable, especially if the system is unique. Without an increased financial incentive, the supplier with the total systems responsibility will be reluctant to gamble on whether he will be paid, when being paid depends in part on the efforts of other suppliers.

In any case, if the customer assigns any degree of systems responsibility to a single supplier, it is quite important that, from the very outset, he play a role in establishing the relationship between the supplier with the systems responsibility and the other contributing suppliers.

Once the design of the system is launched—regardless of who has the overall systems responsibility—there are several additional steps that, if taken at the right time, will help insure successful design, installation, and start-up of the system.

1. The packaging machinery suppliers will establish the package material tolerance range that their equipment can handle successfully, and the package material suppliers will establish the tolerance range of their materials. Some of the listed tolerances, whether they are for packaging machines or for packaging material, are harder to meet, harder to hold, and more important than others. It may prove very worthwhile to have the appropriate packaging machine and packaging materials suppliers meet and discuss their respective concerns about successfully melding machine and material. The customer should participate in this meeting. In addition, when the packaging machine is tested at the packaging machinery supplier's plant, it is beneficial to use a selection of packaging material supplies that covers the entire range of tolerance that is necessary for the job. It will also be worthwhile to have the package materials supplier in attendance at the time of testing.

2. It is certainly less expensive and less time-consuming for both the customer and the packaging machinery supplier to make any necessary modifications of equipment prior to its shipment from the supplier's plant. Thus it pays if the customer visits the supplier's plant at certain stages, to check on the performance and on the construction of the machinery. Perhaps the customer should send, as some of his representatives, the people who will ultimately have to operate the system, since, among other benefits, an individual's understanding of something new increases his receptivity to it, and his ability to make it work well when the time comes.

3. If two packaging machines are to be electrically or mechanically connected for the first time, and each will have a direct effect on the performance of the other, the customer should consider having the two machines connected and tested together by one of his machinery suppliers prior to shipment to his plant.

4. Until the system is proved successful in actual production, the customer should resist any tendency to end the "he" relationship and revert to the "it" relationship with his suppliers. When the appropriate time does arrive for the customer to reposition the responsibilities back to the normal organizational channels in his company, it is worthwhile to work this out with the knowledge and cooperation of the suppliers.

5. Even if the daily charge is slightly higher, and especially if the packaging machine is somewhat new, the customer should consider requesting the packaging machinery supplier to provide installation and start-up by the people who are most familiar with the machine; that is, the people who built it. Those men are best qualified to provide a fast start-up, and, what is at least equally important, they should be able to provide a better quality of informal, at-the-machine training for the customer's operating personnel.

6. During the first year or two that the packaging system is in production operation, the customer may want to consider making service contracts with his packaging machinery suppliers. On occasion, a customer's maintenance and operating personnel are reluctant, and sometimes embarrassed, to ask for outside assistance. This reluctance may be overcome if these people understand that their company has already contracted for, and is obligated to pay for, outside assistance, regardless of whether it is used.

In our discussion of the people, the machines, and the materials that combine to form the packaging system, we have briefly examined the needs and the role of each of these elements in the total system. We have identified certain potential problem areas, and proposed certain steps that can be taken to eliminate the causes of these problems. In conclusion, here are two suggestions that are very important if we are to achieve a greater recognition of, and utilization of, the human element in the packaging system complex.

In order to create the best packaging system for the customer, the suppliers (both material and hardware suppliers) must learn to coordinate their assignments and implement one another's activities (within the proper ethical and legal boundaries) without being forced by the customer to do so on a continuing basis. We must learn how to form one-shot, specific-customer packaging systems alliances.

The customer can best achieve a successful system by identifying those phases of his company's organizational structure, procedures, and routine that should be temporarily bypassed, consolidated, or modified while the packaging system is being created and proved in production. The "single door" concept is most important—a single door policy that encompasses internal corporate activities as well as external relationships with suppliers.

The Packaging Specification

Marshall G. Baldwin

When we examine the technique of writing a specification, we find ourselves involved with a pair of glamorous terms: the two C's of packaging—creation and communication. Without creation there would be nothing to communicate; therefore, creation must precede the actual writing of the specification. The specification is our tool or instrument of communication, and it is this instrument that we shall investigate in some detail.

The writing of a specification can be both an art and a science. The science of packaging must certainly be understood if we are to write a specification, and I believe that some art is required if we are to communicate properly. The packaging engineer is the person who must be able to fulfill both of these requirements. He may be the most creative packaging

MARSHALL G. BALDWIN is supervisor, package engineering, for the Colgate Palmolive Company in New York City.

engineer in the universe, but if he cannot communicate his ideas through the specification, his work may well go for naught. He must be able to communicate with the many different persons, departments, and companies that will actually implement his creative concept.

The Requirements of a Specification

A specification is a statement of the requirements that must be met in order to obtain a given objective. It serves two basic purposes: (1) to communicate the requirements to the parties who will be responsible for their execution and (2) to determine whether the requirements have been met. A specification should be, as the term indicates, "specific." It must be as complete, as clear, as unequivocal, and as concise as possible. Writing more than is necessary into a specification detracts from its value; leaving out essential data is, of course, potentially disastrous.

The People to Whom We Communicate

Packaging specifications are usually thought of in terms of their being directed to a supplier of packaging materials—we often call them "purchase specifications" for this reason. We should, however, stop and ask ourselves, "Is the supplier or manufacturer of packaging materials the only party concerned with this specification?" Certainly not! A whole string of people are involved, and most of them are in the employ of the same company as is the specification writer. Let us consider some of these people and their concern with the package specification.

The purchasing agent. Obviously, the purchasing department is concerned. The buyer must understand the specification in order to discuss it intelligently with the potential supplier. He will not be in a position to evaluate the qualifications of a potential supplier if he does not understand clearly what it is that he is asking the supplier to do.

I do not think that the purchasing agent is unqualified to understand specifications, but I do think that purchasing men, in their anxiety to clear their desks and meet a critical delivery date, are often inclined to hand out orders "with purchase specifications attached," without having fully understood what they have just ordered. Purchasing in itself should not be considered a science; it is a knowledge of the materials that are being bought. So, the specification writer must help the purchasing agent out and write a specification that he can read quickly and understand completely.

If there is any doubt whether you have been able to communicate with the supplier through a written specification, direct communication is required. It is often both desirable and necessary to arrange such a meeting; however, the specification should be written with the idea in mind that this one piece of paper will be the total communication. It may turn out to be just that.

The planning department. The questions of planning what quantity to buy, when delivery is required, and how much inventory to carry are handled in different ways by different companies. For our purposes, we are going to assume that there is one person or one department working closely with the purchasing department in making these decisions. He is not concerned in great detail with the specification, but he must have specific information regarding the number and identity of each component that goes into the finished case unit that is to be offered for sale. We shall discuss the format of the instrument that accomplishes this objective later.

The industrial engineering department. The industrial engineer is the person who must determine production-line requirements in both equipment and personnel. In order to do this, he must have a thorough understanding of the various components, including a knowledge of their physical nature, and how the pieces fit together to make up the finished shipping unit. Outside dimensions, cubage, and weights are going to be of particular concern to him as he determines palletization, truck- and car-loading, and maximum utilization of warehouse space.

Even though he probably was in on the original creative process and had to be consulted on questions regarding machinability and materials handling of the proposed package, do not assume that he remembers all the details—put them all in the specification.

The manufacturing department. Much of the same information that is required by the industrial engineer will be required by the departmental superintendent or the line foreman who will supervise the assembly and filling of the various packaging components. As I have said before, when you write a specification, keep in mind that you are writing for a diverse audience. The men who must read the specification do not necessarily have the technical knowledge, or the familiarity with the technical terminology, that you have. Write so that the specification can be read and understood by all the people who are concerned with its contents. While redundancy is normally not desirable, if necessary say it in two ways, so that you will communicate with everyone who should understand the specification.

The quality control department. Too often, a specification is written only for the benefit of the quality control department. It is too often thought of as a tool designed to help determine whether material is acceptable or rejectable. It is certainly much more than that, however. Criteria for the acceptance or rejection of material must be included, and this is usually done by reference to a number of performance tests that can be run in the quality control department. The individual specification cannot contain all the necessary details about sampling plans and tests; however, it must make reference to the place where this information can be readily obtained. We will discuss how this can be done later.

The cost department. Cost accountants are often interested in reviewing the packaging components for a particular packaged product when they prepare cost estimates. Try to think in terms of what an accountant needs to know when you write the specification.

The legal department. As the government becomes more and more involved with packaging, this department also becomes more and more involved. It is particularly concerned with net weights, marked weights, fill points, statements of ingredients, copy, and, if it is an FDA item, the nature of the packaging materials themselves.

The patent department. It may be necessary for these people to review a specification from time to time. If you write a specification *specifically,* you will probably write it so that the patent department will understand it.

Divisions and subsidiaries. In these days of conglomerates and the internationalization of corporations, you as the packaging engineer are likely to find your counterparts in other divisions or subsidiaries of your company. These people will also benefit from the information that is written into the packaging specification. The original packaging specification should be written so that it can be of service to the packaging men in these related companies. You cannot, obviously, write it in every language on the face of the globe, but you can write it in words that can be easily translated.

If the total specification is to perform all the communication tasks that we have discussed, it must do the following:

1. It must list all components that are to be assembled and tell how many of each are required.
2. It must give adequate instructions as to how to put the components together.
3. It must define company policy, and give general instructions delineating company requirements for a given type of packaging material.
4. It must provide detailed procedures for all performance tests that are required.
5. It must give specific details describing each package component.

We ask you, the packaging engineer, to think beyond the usual concept of a specification that is written only for the complier—think of the other members of your own company who will read and use the specification.

Specification Format

I have several suggestions for a specification format that will help you accomplish this objective. A recent review of the Package Purchase Specification distribution list at Colgate Palmolive Company revealed that we typically send out 39 copies of each of these specifications to five departments, and that there are, in addition, many special distributions on certain items.

In addition to this, it was noted that we typically send out 43 copies of our Standard Packaging Specification to seven departments. Let us discuss this Standard Packaging Specification, and how it fits into the total specification package. This is the part of the specification that is only concerned with intracompany usage.

During the past year, I have done considerable research into the question of how various companies are writing packaging specifications. Our class in packaging specifications at New York University has reviewed the systems employed by a dozen companies, and we have found a remarkable similarity in these systems. The piece of paper that lists the components and describes how to assemble them is variously called the

- Packaging components list.
- Packaging particulars.
- Packaging materials summary.
- Standard packaging specification.
- Packaging order.
- Packaging index.

It varies in size from a 3″ × 5″ card to a 8½″ × 11″ sheet, and it varies in format from a simple list of components to a set of assembly instructions. It is usually filed and identified by the code used to identify the finished product. It always lists all materials to be used, and the number of each required—if the requirements vary between production locations, this detail is spelled out. In many instances, it provides pictorial instruction regarding the proper method of assembling the components. As with all parts of the specification, it refers to its authority and states its reason for revision from its predecessor. This specification should be more than just a list of materials; it should also be a set of instructions. It is the most important part of the total specification as it relates to intracompany communication.

This standard or components specification is not to be confused with a set of assembly specifications issued to a contract packager or to an outside processor who is required to furnish a portion of the total package already assembled. An example of a situation where an outside processor is required is that in which bottles must be packed into reshipper cases. A separate assembly specification is required here to tell the processor how many bottles are to be packed per case, whether they are to be packed upside down or right side up, how to assemble the partitions, what closing method is to be used on the case, and how the bottles are to be palletized. Where such an assembly specification is used, *this assembly, not the individual components,* is listed on the standard packing specification or component list.

Before concluding the subject of specification format, I should like to make two strong recommendations, based on my NYU study. A specification based on performance requirements, rather than on "how to make" instructions, is the goal to be strived for, even though this goal can practically never be fully achieved. The resulting specification will be a compromise of the two methods, emphasizing performance and giving as much latitude as possible to the complier, thereby assuring a maximum of creativity achieved at the lowest cost and offering the best service. If the specification writer has insufficient experience to write a strong performance specification, it will be necessary to include specific directions concerning the use of the materials and the handling methods that are known to work. This is often necessary because the specification writer does not know why a material performs properly. He just knows that it will work, and is afraid to substitute.

My second recommendation concerns the use of a strong "general specification" that covers a class of materials and is supplemented by a simple individual specification, spelling out the details of size, color, weight, and so on that differ from one item to another. This general specification could, as is done in some companies, be presented in the form of a manual. It should cover all general requirements pertinent to a class of materials, such as the basic raw materials, a listing of styles (identified by a code), the procedures for sampling, the procedures for approval of single cavity molds and first-run samples, the procedures for the establishment of acceptable quality levels, and the details of performance test and packing procedures. A new supplier should be urged to study and discuss this general specification.

The advantages of such a system are obvious. A prolonged discussion of this subject is not within the scope of this presentation, as they pertain more to the purchase specification and the intercompany usage of the specification rather than the intracompany use.

In conclusion, my message is: As you write a packaging specification, write it for all the people and all the departments that may have cause to use it. Use simple, direct language that is impossible to misunderstand. Cover everything that is important, but do not cloud the issue with unnecessary detail. Remember, you are communicating with people; if misunderstandings and errors occur as a result of an inability on the part of one person to interpret the specification properly, this should provide a challenge to the specification writer to write more clearly the next time.

Package Design for the Warehouse Function

Kenneth E. Clark

THE main source of contact between a supplier and a distributor is in the buyer-salesman relationship. This relationship provides the main channel of communication, and through this channel are conveyed packaging requirements for warehousing and shipping, problems connected with present packaging, conflicts between the manufacturer and the distributor's packaging needs, and any changes of distribution that require changes in packaging or materials handling.

It is the responsibility of our distribution services group to keep the buyer informed of its needs and problems, and it is the responsibility of the buyer to communicate these thoughts to the salesman. The salesman must, in turn, communicate this

KENNETH E. CLARK is director of grocery purchases for Acme Markets, Inc., in Philadelphia, Pennsylvania.

information to his own manufacturing and distribution services groups, and then follow up on the communication to see that it is acted upon. If there are additional problems at his end, he is also responsible for communicating those problems back to the distributor.

Sales people sometimes look upon a shipping carton as a display unit. They expect that, in addition to meeting its requirements for distribution purposes, it will contain a reorder quantity that is sufficient to keep the shelf stock fresh-looking, and will allow them to maintain turnover quantities. This will keep the product well within shelf life and minimize investment in shelf inventories. The distribution services, on the other hand, think of the shipping carton in terms of packaging that gives adequate protection and is economical to handle. It is understandable that the desires of sales and distribution services are likely to be in conflict with reference to what constitutes an ideal shipping carton.

Outwardly, distribution centers look little different from their counterparts of years ago, and, in spite of numerous announcements made and titles granted to the effect, there still are very few truly automated warehouses. On the other hand, there are many innovations in our distribution facility systems that may require changes in the shipping container specifications.

1. There are now more *high-speed conveyors* in our warehouses, and these conveyors place additional requirements on shipping containers. For example, the product must be able to move easily over the conveyors and stand up under the rough handling and jogging of such systems.

2. Generally our newer warehouses have *higher clear storage areas,* which enable them to accommodate higher stacks. We are now able to stack products 20 feet (five pallets) high.

3. Our experience has shown that the average warehouseman gives the package *less tender loving care* than was true years ago. We find more abuse and more damage, possibly caused by less experienced help but probably caused by less dedicated help.

4. There is no question that warehousing distribution *costs are higher* in almost every area. For one thing, there is a greater investment in buildings and conveyor systems. The largest increase, however, has been in the cost of manpower. This has come about not only because of an increased per-hour labor rate but also because of greatly increased fringe benefits. Unfortunately, productivity has not increased correspondingly in relation to costs.

So much for the changes that have come about at distribution centers—now let us look at what specifications the food chain distribution services want the shipping cartons to meet, in order to facilitate proper and economical handling of the 6,000-plus items that are stocked in an average grocery warehouse.

The weight of the shipping case. Most selection in a warehouse is a hand operation; therefore, the maximum weight should be 60 to 65 pounds. Most cases weigh somewhat less than this.

The shape of the carton. The grocery associations have been discussing a proposal that would make all cartons one of four standard case sizes. We do not feel that this will be the answer, because of the wide physical variations in the stocked products. For example, from small products, such as fingernail polish, to large products, such as paper products, there is too much difference in the density and value of the products for us to think that we will be able to accommodate them properly with only four standard sizes.

What would help, would be to make all cases rectangular. This shape allows a tie or interlock on the pallet patterns. Cases that approach cube shapes are a problem; even if the top layer is tied or looped together, when the pallet is placed in the selection line and the top tape been removed, we find that many packages fall off.

Carton oversizing. Strange as it seems, oversized cartons are currently one of the most common problems. (I am sure, incidentally, that it adds extra cost to the shipping container.) An oversized carton has a tendency to crush, because it does not offer proper support, and this in turn encourages stacks to fall over. It also allows excess play of the products within the carton, which causes abrasion marks on the labels.

Adequate strength and proper dividers. We notice that some cartons have inadequate test strength, improper dividers, and dividers that offer poor vertical support. Dividers often are of a different height from the cartons in which they are fitted, and we also find that they often do not have enough vertical strength. If a carton has the proper dividers and is of the proper case strength, there should be an even distribution of support. In warehouse storage areas, we stock merchandise five pallets (about 20 feet) high. Given these conditions we find that most of our problems are with cartons that contain flexible packaging.

The closure. The manner in which many cartons are closed is inadequate. Most cases are glued, and often the glue is not the correct type, or there is not enough of it, or it is not spotted properly. Cases are sometimes sealed with tape, which is also a problem—especially in frozen foods. After being stored, at a low temperature for a period of time, the tape often comes unglued and curls off. Both conditions, improper glue and improper tape, cause cases or flaps to open. This creates not only a handling problem but also a loss problem through spillage or pilferage.

Cost cutting. Cartons certainly should be made as economically as possible, but they still should serve their need. Some manufacturers have cut costs by making too wide a gap between the flaps. Our selectors use these gaps as handles, and, because of either inadequate board strength or poor gluing, unintentionally rip open the case and spill the contents when they pick it up. This could be corrected by reducing the gap, by using stronger board, or by using adequate glue. As I mentioned before, the cases quite often have inadequate board test weight, and they do not give adequate protection.

Identification. The larger, nationally advertised manufacturers generally do have good identification on their cartons.

They identify all four sides, the products are color-coded where desirable, and the lettering is large enough to read at the required distances. The manufacturers of private-label, regional, or packer-label brands are the ones who generally do a poor job of identification. We sometimes find that there is identification on only one side of the carton, and it is quite often a poor stenciling job at that.

We need adequate lettering that is readable at 20 feet on *at least* the two ends. Color coding is preferable when there is a difference in styles or colors of a given product. The carton identification should clearly describe the contents. For example, it should indicate whether the contents are multi-packed or single units, and if there are several different colors available for any one item, the identification should indicate the color of the items in that carton. An important point to remember in regard to coding is to avoid using a color coding that is different from the color of the items inside.

Unitization. A pallet unit should hold together not only during shipping but also during the selection period when the unit sits on the floor or in a pallet rack. This does not mean that the cases should be spot-glued together in such a manner that it takes a crowbar to separate them. It does mean that the cases should be designed so that there is a tie-together without excessive overhang. In receiving, we sometimes find that when one pallet unit contains two or more products, the cartons are sometimes spot-glued together and are quite difficult to separate.

A major problem is caused by manufacturers who do not use a uniform pallet or layer pattern on the same item. This occurs when the items are manufactured at different places, and it creates a great deal of confusion during stock taking.

In summary, these are the characteristics of shipping cartons that are desirable from the point of view of a food chain distribution service. The shape of a case should be such that the layer pattern will tie the cases together and prevent them from falling off the pallet. The weight should be such that a man can easily pick it up during selection. The board strength intend to make our voice heard, on the ground that in the American economy every creditable product is entitled to an equal opportunity to compete. The paper industry is willing to take its chances in any market if the competing products are allowed to sell on their merits, rather than on the basis of what we consider to be lopsided legislation.

But looking at the packaging picture as a whole, I think it s unmistakable that competition from plastics has helped, not urt, the paper industry. It has stimulated our manufacturers develop improved materials and highly salable new packging systems. When a new development in plastic packaging s appeared, such as the shrink-pack or lamination, the per industry has been quick to adjust. One example is the ltiwall bag. By combining polyethylene film with kraft er, we now make a bag that performs certain functions ter than any all-paper or all-plastic bag. In short, paper plastics may be competitors, but they are also allies. ther they can do jobs for the consumer that neither can parately. It is significant that in 1969 almost two-thirds of ollar value of all plastics used in packaging was cond by the paper industry.

t long ago a packaging man said that, even though and paperboard are essential, as packaging materials ck glamour. His attitude struck me as being a little like an affectionate husband toward a wife of long standloves her, he needs her, but he takes her for granted; en a sexy newcomer catches his eye, it's not surprising houghts stray in her direction. The fact remains, that paper lately has been producing some pretty packages of its own—and it continues to be the le mainstay of the packaging industry.

w much of a mainstay paper is comes through in a tistic: During the 1960s, about 95 percent of all red goods moving in the U.S. economy were corrugated packaging. Corrugated is of course an important factor in our production picture. Last gated produced $3.5 billion in sales. That is nearly

of the carton should enable it to be moved safely on conveyor systems, to be stacked 20 feet high, and to withstand normal abuse, not only from the machines, but also from the men who handle it.

The closure must be adequate, and the identification must be complete on at least two ends in clear letters that can be read at least 20 feet away. As for product protection, we expect the case to protect the product against crushing or breakage; however, we also expect it to protect the product against moisture or infestation. It is unfortunate that the most perishable products frequently seem to be the least well protected by their shipping container. The baling of dog food and of flour are two cases in point.

The principal source of communications between our distribution services department and a vendor is through our buyers. Likewise, our principal communication with a supplier's manufacturing or distribution service is through its salesmen. Let us be sure that the lines of communication are kept open and used. It is only through this free communication that we will be able to pass along suggestions, make known our problems, seek corrections for those problems, and search for improvement. There will always be conflicts among the manufacturing, distribution services, and sales people, but with good communication we can determine what these problem areas are and seek satisfactory solutions.

Paper Packaging: The Outlook for the Seventies

Edwin A. Locke, Jr.

THE growth of paper and paperboard packaging has been phenomenal in recent years. One good measure of the gain is the output of containerboard, the raw material of corrugated board. In the ten years from 1960 through 1969, containerboard production nearly doubled, rising from about 8.5 million to 16 million tons. Few other commodities in the American market have shown a rise of such proportions. And an equally significant fact is that the total value of paper and paperboard consumed in packaging rose from about $4.5 billion in 1958 to over $8 billion in 1969—again nearly double. It is worth bearing in mind that only a small fraction of that increase was due to rising prices. Between 1960 and 1969 the

EDWIN A. LOCKE, JR. is president of the American Paper Institute in New York City.

Wholesale Price Index for pulp, paper, and pap
rose only 7.2 percent, an astonishingly small pric
the 1960s. Yet even with prices that had advan
paper and paperboard in 1969 represented 44 p
dollar value of all packaging materials, includi
tics, glass, metals, and everything else used in t

These statistics are cited in order to make t
we enter the 1970s, the position of the pape
packaging market is strong and encourag
slowdown in the economy may make it temp
top our 1969 figures, and we shall be hard p
rate of growth of our packaging sales in
comparable to that of the past ten, but ev
ative projections point to substantial g
government forecast indicates that six
tainerboard production will have rise
1969 tonnage figure and will be more t

Occasionally, we are asked by the
the paper industry is not disturbe
plastics in packaging. Our answer—
—is that we are much more stimula
In my own opinion, nothing bett
the paper industry than the deve
petitive material. Not many yea
turers had good reason to feel
market—if you'll forgive the
today, even with business at a
complacency in the industry.

It is perfectly true that
competition from plastics ha
used to package meats in
plastic and foam meat tra
places. But even here it i
concerns us as the move
competition. Our objec
troubles us is the effort
against pulp trays. W

45 percent of our industry's total packaging sales, and 20 percent of the total packaging market. Now and then, fears have been expressed that plastic in its various forms was going to cut deeply into corrugated tonnage, and in 1969 the Fibre Box Association retained Arthur D. Little, Inc., to make a study of the position of corrugated in packaging. Although the report has not yet been made public, it has in it much to encourage our industry, and it emphasizes the big additional opportunities now opening up for producers of corrugated.

There is good reason to think that two recent developments in corrugated packaging are likely to prove of special importance in the 1970s: the multiuse corrugated box, and the super-strong lightweight container for air freight. The time is not far off, I believe, when the typical corrugated package will be stronger and lighter than in the past and will serve the shipper not only in distribution but in marketing, in merchandising, in the storage of goods, and in inventory control.

The Paper Industry and the Environment

In everything I have said up to this point, I have restricted myself to the conditions of the market as we have known it. But we all realize that a number of these conditions, and some mighty important ones, are going to change, and change sharply, in the near future. The problem of environmental protection has become of paramount importance for all of us; a nationwide effort has recently been undertaken to impress on the American people the importance of swift action to cope with environmental pollution. Certain adaptations will have to be made by people who produce and consume packaging.

Earlier this year, in his message on the environment, President Nixon stressed the problem of packaging waste, saying: "New packaging methods, using materials which do not degrade and cannot be easily burned, create difficult new disposal problems. . . . I have ordered a redirection of research under the Solid Waste Disposal Act to place greater emphasis on techniques for recycling materials and on development and use of packaging and other materials which will degrade after use—that is, which will become temporary rather than permanent waste."

That is direct and to the point. In effect, the packaging industry has been put on notice to adapt its operations to the production of packages that can be easily disposed of or recycled. The President did not tell us how much or how little time we had to make this big adjustment, but if there is any doubt on this point, it is being rapidly cleared up by the Congress. Senator Gaylord Nelson of Wisconsin recently introduced a bill called the Packaging Pollution Control Act of 1970, part of which calls on the Secretary of the Treasury to put into effect disposal charges for all packaging that may result in solid waste or lowered environmental quality. That schedule, under the bill, would go into effect not later than June 30, 1971. Furthermore, the Secretary of Health, Education and Welfare would be empowered to make sure "that all packaging will be returned, reused or recycled into the economy, and will be made of such composition and construction as to be degradable." For violators of the code that is to be established, there would be penalties ranging from $1,000 to $5,000 per day of violation.

The penalty provisions of this bill might prove extremely difficult and costly for the government to enforce, and in my view do not represent the best means of achieving the objectives of the bill; but whatever the fate of this particular legislation, it shows the general direction in which the country is moving. Not only government but private financial interests are thinking along these lines. Early this year, the executive committees of the three leading bankers' associations of the state of Maine approved a Bankers' Pollution Code, calling on banks to consider environmental factors when deciding whether or not to lend money to industry. The old saying, "as Maine goes, so goes the nation," may not always be true in politics, but I believe it is very likely to be borne out in this instance.

Any industry that requires large amounts of capital for investment but does not do its full share to abate pollution is liable to find itself in the position of the man who was recently brought before the Chief Justice of the Federal Tax Court on a charge of tax delinquency. This defendant made an emotional plea, and wound up by saying, "As God is my judge, I do not owe this tax." The Chief Justice replied, "He's not, I am, you do." In the years ahead, bankers may well be saying things like that to environmental delinquents.

Adjusting to the new requirements will not be easy, but the pulp and paper companies fully recognize the need for action, and are on the move. The industry has already done more to combat pollution than many people realize. The National Council of the Paper Industry for Air and Stream Pollution has roots going back as far as 1922, and our industry has spent more than $800 million on the tremendous job of environmental protection, at a rate that has increased rapidly in recent years.

Solid waste similarly has long been a matter of concern to us. In 1969 we recycled some 20 percent of the total amount of paper used in this country—about 11.5 million tons of waste paper. Currently, we are running pilot programs to find more effective ways of collecting and recycling newsprint and various types of paper packaging. But whatever we have accomplished so far, the effort clearly has to be accelerated.

We are not waiting to be told what needs to be done. Package design criteria are already changing in the direction of greater disposability. Inevitably, the technological and economic difficulties in the prevention of packaging waste are formidable. In this industry, an ounce of pollution may require a pound of cure. And the cure demands that we make more progress with respect to substances that prevent boxes or wrappings from being recycled. More research is especially needed on ways to make highly functional packages more easily degradable or more easily recyclable.

Many recent innovations have of course added greatly to the versatility of paper packaging, and to lose that versatility would be an injury to our customers and ourselves. But the hard fact is that the better the package is from the customers' standpoint, the worse it sometimes is from a disposal standpoint. The big job ahead of us is to find ways to reconcile these two objectives—to produce packages that serve our customers as well as do those that we are making today, and to make them capable of being readily collected and recycled. This is a major enterprise calling for intensive research and creative imagination, and above all for the cooperation and assistance of designers and users of packaging at all levels. It strikes me that there may well be an urgent demand for a new kind of engineer—the disposability engineer—and I hope that technical schools will soon begin to focus more attention on the training of specialists in this field.

Another problem that demands prompt attention is to find ways to cut down the amount of used packaging to be disposed of by consumers without reducing the usefulness of the package for the protection of goods and for self-service merchandising. The solution to this problem, I believe, must rest largely with the package designer. There is at times a natural temptation to use a little extra something in order to achieve a more attractive or convenient package. But the fight to protect the environment—which means in the long run to protect ourselves—makes it essential that from now on, in designing every package, we give a high priority to the question of disposability.

Even if the package can be made more disposable and kept to minimum weight, we still have to devise more effective methods of collecting packaging waste. For example, waste newsprint is now being collected through a number of channels, including Boy Scouts and other public-spirited groups, who sell it to wastepaper dealers. The dealers in turn sell it to mills that have the necessary facilities for recycling. In some cases, the old newspapers are de-inked and recycled into newsprint, but most mills convert them into what the trade calls combination paperboard. Through this system in recent years, an average of 20–25 percent of the newsprint consumed in this country has been recovered—in 1969, about 2.5 million tons.

We came to the conclusion that the percentage could be raised by combining better collection techniques with public education, and this opinion is being justified by a pilot test now being made in Madison, Wisconsin. In the test area, specially designed sanitation trucks are picking up at least 40 percent of newspapers going into the area—about twice the national average—and the city is selling the salvaged newsprint to dealers who return it to mills for recycling.

Improved methods of collection for corrugated are also urgently needed. We are currently studying methods to obtain greater cooperation from supermarkets and department stores in collapsing and baling corrugated boxes after their contents have been stacked on the store's shelves. At the present time, some 24 percent of used corrugated boxes are being recycled, and we hope to drive that figure up sharply. We are also giving close attention to the possibility of obtaining cooperation from offices, stores, and housewives in sorting the mixed papers that they throw out, as a way of increasing salvage.

It is not an oversimplification to say that what the country most needs in order to cope with the solid waste problem is a combination of the right technology and the right motivation. And I would add that the word "motivation" (or, if you prefer, "incentive") cannot be too heavily stressed. The entire success of the government's effort to put an end to solid waste pollution will, in my opinion, depend on the degree to which it is successful in persuading all elements in our society to cooperate. This is also the view of government specialists.

Last year, a comprehensive study was made by the Department of Health, Education and Welfare, entitled "The Role of Packaging in Solid Waste Management, 1966–1976." The study says clearly that "incentive-type mechanisms appear to have the best chance of being effective in attempts to improve salvage." The study goes on to propose two possible incentive measures: price supports for secondary materials collected in salvage, and tax credits for scrap processors and industries using secondary materials. These credits would be designed to stimulate investment in plants for salvage purposes. The case made by the government study is powerful and convincing, and certainly deserves careful consideration by the administration and Congress.

One thing is certain: The time is at hand for an all-out combined attack on the packaging waste problem by all the groups concerned, in cooperation with government. Major readjustments are extremely difficult for industry, and especially for industries with big capital investments, like pulp and paper. We know there will have to be costly changes in design and production, to say nothing of planning and administrative procedures. Familiar patterns of operation based solely on market requirements and preferences will have to be modified as never before to take into account the necessities of waste disposal. But the job needs to be done, and, knowing the paper and packaging industries, I predict that it will be done successfully. A few years from now, when we review the work of the early 1970s, I believe that we shall look back with pride on what we have achieved for the environment, and that we shall be able to look forward with confidence to even greater marketing opportunities than before.

Solid Waste Management in Packaging

Eric B. Outwater

On "Earth Day" 1970, people from coast to coast demonstrated against further degradation of our environment —and, for possibly the first time, they had the full endorsement of federal, state, and local governments. We are, in their words, in the throes of an "ecocatastrophe." It is not my intention to add my voice to the chorus of protests against packaging in general, and one-trip packaging in particular, but rather to tell you what the protestors are saying, why they are saying it, what communities are doing in the area of waste disposal, and, most important, what positive steps the packaging industry can take.

Most of us know the story about the urban mother who tells her child, "Hurry up and eat your soup, dear, before it

ERIC B. OUTWATER is a consultant to the Foundation for Responsible Conservation of Our Environment, Inc., in New York City.

gets dirty." A good deal of this dirt is caused by the inadequate degradation of substantial amounts of packaging.

Although the major thrust of the demonstrators is at air pollution (cars and utilities) and water pollution (detergents and sewers), garbage and litter (synonymous with packaging in most people's minds) are the demonstrators' arch-enemies. Let us look at some of the statements and instructions that they have been issuing, the gist of which can be summarized as: "The packaging you take home today becomes trash tomorrow."

Their main contention is that trash disposal, in addition to being costly, can be highly dangerous; that burning discarded packaging in building incinerators and city dumps contributes to air pollution. The burning of paper gives off carbon monoxide and particulates, and the burning of the new polyvinyl chloride (plastic) packaging materials produces hydrochloric acid, a very toxic substance. To combat this pollution, they suggest that consumers bring their own, nondisposable containers to stores and markets, fill them with merchandise, and leave the commercial packaging in the store for disposal or recycling. The emphasis, they say, should be on recycling.

There is no sense in our saying that really it is the consumer who wants convenience packaging, and that we are not responsible for the consequences. While it is obvious that everyone would like to have convenience packaging without the responsibility of disposing of it, or of depositing it in an unobtrusive area and in more easily disposable and degradable form, the packaging industry must carry part of the burden of creating a new kind of packaging that is both convenient and readily disposable and degradable.

Three years ago, William May, chairman of American Can Company, said, "Let's solve the problem of pollution within the packaging family." One year later, I repeated Mr. May's remarks, adding my own concern at the fact that there had still been no visible attempt to implement his suggestions, and my feeling that some day there might be an Earth Day and that we would not be ready. Now Earth Day has come and gone, and, with few exceptions, we seem to have given up on an industrywide basis. On an individual company basis, it is another story:

1. International Paper has conducted studies on dumping packaging waste at sea and has actually made such dumps in the ocean and studied the results. It is interesting to note that the New York Regional Plan Association has also suggested ocean dumping as the ultimate site for cities that border on the ocean, but has noted that a great deal of work has to be done to prevent the materials from floating back to the beaches. Far more thorough studies have to be conducted to learn the effect of ocean dumping on the ecological cycle in the ocean.
2. Numerous companies, particularly in the soft drink industry, are computing disposal costs into the total cost of their product and taking a good, hard look at the effect of PVC on incinerators. When PVC is burned, it turns out appreciable amounts of hydrogen chloride, which not only contributes to serious air pollution but corrodes metal incinerator surfaces.
3. One electric products company has switched from corrugated trays and inserts for its TV tubes to reusable trays.
4. A glass plate manufacturer has switched from wirebound boxes to steel racks.
5. A great deal of work is being done within marketing limitations to reduce the number of square inches of each package, not only to reduce cost, but also to reduce the amount of waste when the package has been discarded.
6. A Swedish firm is studying the degradability of various packaging materials when they have been buried in the ground.

Many of the remarks that have come out of the packaging industry have been negative. William May stated in a speech before the New York Board of Trade this January that "we cannot expect corporations to undertake environment protection on a voluntary basis, because of the tremendous pressures of competition." Certainly, he is correct if he speaks for one segment of the industry working alone; no one would expect a firm to give up making beer cans or to engage unilaterally in costly research into degradability.

But when we abdicate responsibility for a problem that is becoming increasingly important in the public mind—and let us admit that litter is visible, packaging is durable and bulky, and the public pressure to alleviate garbage disposal problems is building—the government will gladly fill the vacuum. First, money is made available for research and demonstration grants, although these have been extremely limited, and seem to be decreasing rather than increasing. Then, financing is provided for methods of disposal such as incinerating, compacting, and composting. But solid waste is at the bottom of the priority list, after sewage plants and air pollution control.

Finally, if these efforts are not effective, laws are proposed that affect the source. About 50 proposals have been put forth in as many legislatures to limit packaging, but so far, few have gotten out of committee. Still, even if we in the industry tend to resent governmental control, we must recognize that we have a moral commitment to future generations to find a sane balance between technology and the world in which it exists —not to waste our natural resources by filling the ground with plastic, aluminum, and glass.

Let us take fair warning: If demonstrations can keep a president from running for another term, they can certainly put a crimp in industry, particularly during an election year in which conservation is a key issue. Public clamor often runs from apathy to hysteria and back to apathy again in three years—but don't count on it this time. The raw materials producer, the converter, and the manufacturer whose products are contained in packaging must do something positive *now*.

In a recent issue, *Package Engineering* stated that the

leadership role of the federal government in solid waste management is now clear:

> Probably every important newspaper in the country carried at least the high spots of President Nixon's February message in which he urged research as well as action on package disposability. Less well known is the earlier testimony of Robert H. Finch, Secretary of Health, Education and Welfare. His testimony before the Senate subcommittee on air and water pollution was a prelude to what is becoming the position of the government. His staff members are now using his words as their guide:
>
> "The burden should fall primarily on those who produce wastes, to provide incentives to eliminate, reduce, or reclaim these wastes," he stated. He also put a finger on who would pay the cost: "For solid wastes, in particular, we need to study the feasibility of placing the responsibility on financing the management of wastes on those who produce them."

The secretary also turned his guns on packaging: "We have an increasingly severe problem of container disposal. We need to study ways of removing bottles, cans, and other containers from the category of 'social cost.' It may be 'efficient' for industry to use these containers *only* because the efficiency formula *excludes* the cost to society of disposing of indestructible, nonreturnable containers." (The emphasis was his.) Mr. Finch said further that the separation and reclamation of such containers would clearly benefit society but that the nonreturnable container does not.

Mr. Charles C. Johnson, Jr., Administrator of Environmental Health Services, thinks that people using a package should help pay the cost of getting rid of the empty package. He foresees the development of ways to build the charge for disposing of the package into the cost of the product. In the interim, officials such as Mr. Johnson are making an all-out plea for help from industry. "What does industry—particularly, packaging—have to offer? On what basis can government and industry get together?" Mr. Johnson asks, making it abundantly clear that the welcome mat is out to anyone in industry with helpful thoughts on this massive challenge.

Waste collection and disposal now cost the American public $6 billion per year, and the price keeps escalating. The word "disposal" is an unfortunate choice of words; "deposit" would be more apt. More than 80 percent of the solid waste in the United States goes into the ground pretty much in the form in which we discard it. The alternate methods of disposal are also far from ideal. The most common is to dump into an open pit, where air-polluting combustion takes place. Then leaching by rain pollutes the water table, and vermin infestation adds another unpleasant dimension to the problem. The unsavory sight of such an area during its use, and the uncertain qualities of its future use as a site for home building or as a recreational area, are hardly compensated for by the cheapness of the method.

When waste is placed into the ground, and a residue remains from all methods of disposal, a technique called sanitary landfill should be used. Mechanical equipment not only compresses the waste to reduce its volume but also covers it daily with a layer of soil so that any burning is inhibited and exposure of the waste is at a minimum.

We tend to forget the fact that many items of solid waste are virtually indestructible, even after their deposit in the ground. A plastic bottle buried today will be in nearly the same condition ten or even a hundred centuries later. Actually, the degradability even of a newspaper in a landfill is dependent on how much moisture is present in the ground. In the relatively dry Southwest, newspapers buried 20 years ago are still quite legible.

It is rapidly becoming more evident that the combination of on-site and municipal incineration into a systems approach to solid waste disposal may be the best means to deal with refuse accumulation in our urban areas. This method also makes certain valuable auxiliary functions possible, such as the production of power, heat, steam, or cooling. A modern incinerator is a process furnace with auxiliaries and controls for burning the combustible matter of refuse and for producing a low-volume residue of noncombustibles. If a mixture of average office waste weighs 1,000 pounds and occupies 200 cubic feet before it is put into an incinerator, the residue after combustion will weigh less than 50 pounds and will occupy about one cubic foot.

It is interesting to note that, in the United States, about 13 percent of waste is accounted for by packaging, which has a per capita generation rate of 550 pounds per year, or a total of 60 million tons. The statistics for *total* waste (including packaging) range up to 350 billion tons per year, which works out to the fact that Americans produce 100 tons of waste for every man, woman, and child in the *world.* (This figure is arrived at by adding such items as dirty bath water.) Let us look at a breakdown of the packaging waste statistics in terms of individual raw materials and their percentages of inert residue that must still be disposed of after incineration:

	Percentage of Total	*Percentage of Inert Residue*
Paperboard	34.0	3.5
Other paper	22.0	4.5
Metals	14.5	37.0
Glass	18.0	52.0
Wood	7.0	0.5
Plastic film	1.6	0.3
All other plastics	1.4	2.0
Textiles	0.5	0.01

If we are going to recycle these wastes, we must have an economic reason for doing it. If we are going to recycle glass in the form of cullet, it must be economically worthwhile. Newspapers returned to the pulping process must be no more expensive than virgin pulp. Unfortunately, there are few opportunities, today, to recycle anything on the basis of straight economics. A few exceptions are homogeneous waste materials produced by a plant, which are usually still at the plant and can be replaced into the production cycle; some corrugated containers; clean newspapers, which can be returned for repulping and made into newsback board and molded pulp; and, finally, heat energy from the incinerator process, which can be used to produce steam and electricity. The Btu. of waste charged into incinerators has about one-third the value of coal. This value, of course, decreases as more paper is recycled.

Since one-trip packages are the focal point of discontent these days, let us consider the possibilities of returnable glass. A truism about litter is that "pollution is a matter of choice." In a test conducted last year in Alabama, ten-cent deposits were placed on 1 million bottles. Almost all the bottles were lost within 12 months; people didn't value the deposit enough to bring back the bottles.

Other methods of disposal have been tried, but most have severe limitations. For example, *compacting* reduces volume, but if attempts are made to ship the compacted items, the waste expands to part of its former volume. *Composting* is another experiment that has not worked out too well. Compost, at best (when most of the glass and steel is removed), is still the product of an expensive four- to five-day composting cycle. It has been produced, in the past, in areas pretty far removed from the farms where soil enrichers might be needed, but I suggest that small, more sophisticated digesting systems are still a possibility in small farm communities.

Packaging by its very nature is a combination of materials: Cans are often aluminum, tin, and steel; bottles are glass of various chemical analyses, often incompatible with the aluminum ring left by the pilfer-proof cap; milk cartons are paper and plastic. When incinerated, this plastic in many instances generates a corrosive atmosphere that can close down an incinerator. With all these negatives, packaging and solid waste still do not represent a crisis, only a problem, but the crisis will be upon us. An incinerator completed with federal help takes five years to finish, and in the meantime, the pressures from the population explosion continue to add litter to our woods, water, and streets.

The coming crisis in disposal is complicated by the arbitrary political boundaries between communities. Where one area has a high population density, another may have land available for disposal. Getting the two communities to agree on a mutually acceptable arrangement has often proved impossible. Compounding the difficulty is the fact that physical boundaries of watersheds, airsheds, and disposal areas are not likely to coincide.

Leonard S. Wegman, a consulting engineer specializing in municipal solid waste, testified before the Senate public works air and water pollution subcommittee on the proposed Resources Recovery Act of 1969 (S.2005), stating that S.2005 "has commendable goals" but "will fall far short of the cities' needs," because residential and commercial discards are increasing at an enormous rate—paper consumption has more than doubled in the past 20 years, and plastics consumption has jumped 759 percent in the same period. He offered a detailed version of a concept proposed by the solid wastes committee of the New York Board of Trade council on the environment, which was designed to provide a trust fund of about $3 billion annually for local waste disposal projects, outlining

1. That a national disposal fee of one cent per pound would be imposed at the manufacturing level on all goods and on their packaging.
2. That the equity of such a fee system would be that each article—which would ultimately require collection and disposal services as solid wastes—directly and by itself would generate the funds for such services.
3. That the manufacturer who last prepared the article for consumer use would pay the one cent per pound.
4. That such a weight charge would encourage manufacturers to produce lighter and less complex containers.

5. That the manufacturer would pay his assessment periodically into a federal government trust fund for all the articles he produced, on a weight basis.
6. That the government would routinely distribute the funds as federal supplements to all municipalities, per capita, that performed responsible tasks for refuse collection and disposal.
7. That the local community would continue to decide what form of refuse collection and disposal best suited its own needs.
8. That discards, such as commercial waste or urban demolition, that had not issued as consumer goods from an established point of manufacture would have to be paid for by the source that sent it to collection and disposal.
9. That the national disposal fee would generate approximately $3 billion annually at present consumption levels.
10. That the national disposal fee would enable federal and state governments to mandate the highest standards of environmental protection in the handling of solid wastes—because the cities, for the first time, would have a source of money to meet these mandates.
11. That the monies collected and then allocated to the municipalities would be treated as supplementary funds and not as substitutes for budgetary allocations for refuse services.
12. That each municipality would be required to furnish the administrative agency with a working blueprint of how it would apply the funds to be received from the national disposal fee.
13. That whether or not a municipality provided full or partial collection and disposal services, the municipality would receive all monies to which it was entitled on a per capita basis, with 60 percent to be apportioned to collection operations and 40 percent to disposal.

14. That the federal administrative agency would establish parameters for performance and for appraising how a municipality had used federal supplements to upgrade its solid waste services.

What I suggest is this—that around the nucleus of the Materials Disposal Research Council we consider an organization called, perhaps, "Packaging for People." It could be headed by a director under contract, and financed by an adequate assessment based on each package maker's production. This assessment would be enforced by a symbol placed on each package, similar to the union label, which would be available to those who paid their assessment.

What will "Packaging for People" consist of? I suggest a permanent staff that works with the director, consisting of economists, sanitation engineers, R&D materials and packaging experts, advertising and public relations men, experts in government relations, and an international coordinator. These people should institute and coordinate programs and publicity for the long-range good of the packaging industry in the area of ecology.

We need programs to stimulate recycling. The reuse of materials can, of course, be in the form of energy from burning or the reclaiming of homogeneous waste such as aluminum or paper. More important programs might be formulated to stimulate sorting and separating waste at the consumer level. Also needed are programs to stimulate the public and markets to help with litter and waste problems by crushing containers, returning multitrip packaging like glass, and supporting litter preventive programs such as Keep America Beautiful. Other programs might be

1. Programs to stimulate and guide meaningful research into methods to improve the degradability of packaging.
2. Research in methods of reducing the cost of pickup and transportation of solid waste.
3. Programs to improve the reduction of volume through

incineration, compacting, and the proper maintenance of landfill area.

4. An information and speaker service for all those interested in packaging and package waste programs.
5. A public relations program to acquaint the public with the advantages of one-trip packaging.
6. A service to the packaging industry in the area of anticipated legislation.
7. The dissemination of this information within the industry.
8. The coordination of worldwide techniques, programs, and, most important, research.
9. Cost research, in terms of the public, industry, and government.

It must be apparent that James Reston's recent statement in *The New York Times* that "the old optimistic illusion that we can do anything we want is giving way to doubt, even to a new pessimism" is confirmed by our seeming plunge into a chaos of cause and effect in the constant changes in our human environment. We must hope that man's massive interventions will not, in William Faulkner's words, "turn the earth into a howling waste from which he would be the first to vanish."

I am not only an ecologist but also a packaging man who has worked most of his adult life in this industry, and I believe that as the public protest grows, we should be ready to defend what is truly worthwhile by a positive, not a defensive, approach. We have done so much for the "quality of life" through a vast variety of products brought to the home in convenient, labor-saving, sanitary, and inexpensive form. Let us not be known as the industry that allowed this quality of life to be destroyed through apathy or hysteria.

Packaging for Foreign Markets: Laws and Customs

Joseph M. Callewaert

THE problem of packaging for foreign markets is an aspect of international trade that is ordinarily neglected in the United States, even by packaging people. Perhaps this is because it is such a complicated subject that it has been able to capture the interest of only a few specialists. Nevertheless, it is an important area of business, and should be given as much consideration as is given to any other legal matter in an organization.

U.S. Export Regulations

Obviously, when you send goods to a foreign country, you have to comply with United States regulations. They are usually rather simple, but if you overlook any of them, you

JOSEPH M. CALLEWAERT is president of Business Services International in Suffern, New York.

may find yourself in trouble. First of all, you will need either a validated license or a general license, depending on what you are shipping where. The vast majority of goods can move to the free world under appropriate general licenses. Validated licenses are necessary for certain types of strategic goods, regardless of destination, and for all but a selected list of items intended for the Soviet bloc countries. There is a total embargo on United States trade either directly or indirectly to Red China, North Korea, and North Vietnam, and only non-subsidized foods and pharmaceutical products may be sent to Cuba. If you need a validated license, apply to the Office of Export Control, Bureau of International Commerce, on Form FC-419.

You must file a Shipper's Export Declaration with the district director for all commercial shipments. This declaration is used for compiling official U.S. export statistics and for enforcing U.S. export control laws. Include a specified destination control statement on the Shipper's Export Declaration, the bill of lading, and the commercial invoice, and be certain that all export control documents relating to a shipment under a validated license are in agreement as to exporter, commodity, intermediate consignee, ultimate consignee, and country of ultimate destination. Do not deposit goods on a pier, dock, or other place of loading for export until the Shipper's Export Declaration has been authenticated by the district director of customs. Furthermore, if you know or intend that the exported commodities are to be re-exported, it is essential that the Office of Export Control give prior approval, either through the establishment of a general license or by specific approval of the transaction.

Because of the technical aspects and complexity of export control operations, most firms with a large volume of export business utilize the service of freight forwarders or customs brokers. These forwarders and brokers have personnel available who are experts in the field, as well as correspondents in the country to which the products are being exported.

Foreign Customs Regulations

The first place to go for information on foreign customs regulations is, of course, to your freight forwarders or customs brokers. But be sure you understand them completely; if you are in doubt, by all means consult additional sources of information. For example, try the U.S. commercial attaché in the country to which you are exporting. He is usually very competent, friendly, and willing to help, and knows many people in the country. He can be helpful in avoiding mistakes and in cutting red tape. Also consult the Bureau of International Commerce in Washington, D.C. This bureau has international trade specialists whose job it is to help in any way they can. Simply specify the country or countries of interest to you, and you will get the right person and the right information.

Other sources of information are the magazines *International Commerce, Overseas Business Reports,* and *Economic Trends Reports,* all of which describe foreign import tariff systems. You can subscribe by writing to the Office of Publications and Information, U.S. Department of Commerce. If you wish to have more detailed information, you should subscribe to or consult the *International Customs Journals,* published by the International Customs Tariff Bureau in Brussels (Belgium), which give complete English-language translations of foreign tariff schedules. Tariff structures and rates shown in the *International Customs Journals* are subject to change, and supplements are issued from time to time. However, to be accurately informed, you should have the original, native-language publication (and a few linguists in your company). Here again, your customs broker should be of help.

It goes without saying that another source is your client. If he buys your products, he will certainly want to get them as soon as possible and advise you on the tariffs in his country. Or you can check with the Trade Expansion Committee of the American Steamship Lines in Washington, D.C.; the International Committee of the U.S. Chamber of Commerce, also in

Washington, D.C.; and any banks that maintain offices abroad.

I do not know of one independent country in the world—and there are about 140 of them—that does not have customs regulations. Almost the first move for a newly independent country is to create a customs organization together with a new flag, a national anthem, and an airline. Not so long ago, in its issue of November 10, 1969, *International Commerce* carried a summary of the 140 tariff systems in the world. It does not make interesting reading, but it is impressive—and complicated. Let us try to develop a general overview about the important aspects common to all customs regulations.

Classification system. When dealing with the customs regulations of a foreign country, check its classification system. It might adhere to the Brussels Tariff Nomenclature (BTN) or to the U.N. Standard International Trade Classification (SITC). Some countries, like Canada and the United States (TSUSA), have their own classification systems.

The classification system is important because it establishes the rate of duty. And it can make a difference! For example, nonwoven fabrics are a new product without a proper classification, and can thus be shipped to certain countries (including the United States) either as textiles or as paper. Obviously, the exporter will choose the lowest rate. However, a false declaration to take advantage of a lower rate should never be allowed.

Range of duty. The range of duty is usually indicated by a percentage of CIF value. To this, sometimes, a certain amount per unit of weight is added (francs per kilo or cents per pound). Certain countries have one-, two-, or three-column duties, and each of these countries can be a low-, medium- or high-tariff country. Canada, for example, is mildly protective and a three-column tariff country. Goods from the United States are accorded most-favored-nation or middle rates. The lowest or preferential rates are reserved for the British Commonwealth countries; the highest or general rates apply to countries with which Canada has no treaty or trade agreement. Quite a few countries have preferential tariff systems; that is, systems that apply lower tariffs to imports from specified countries.

Special features. Look for any unusual aspects of a country's tariff system that may be of importance, such as special duty-free treatment, surtaxes on imported goods, taxes per package, and "other taxes." For example, some products not manufactured in Australia may enter there duty-free; the Barbados charges 13 cents (U.S. currency) per imported package; and Panama collects a surtax of 3.5 percent ad valorem (FOB value) on all imports. Some countries—Peru, Chile, Ecuador, and even Great Britain—request prior import deposits and monetary stabilization surcharges, which range from 5 to 10,000 percent of CIF value.

Brazil has an interesting example of "other taxes." In that country, import tariff rates are mostly ad valorem and range up to 205 percent. Most imports are subject to a port improvement tax of 2 percent CIF value; a merchant marine tax, 20 percent of ocean freight; and an industrialized products tax, varying from 4 to 30 percent of duty-paid value.

The Kennedy Round. Each country that participated in the Kennedy Round of tariff negotiations will grant tariff reductions. They are scheduled to come into effect in stages, and the last stage is to be implemented January 1, 1972.

Foreign Packaging Regulations

Basically, a package is a container made of one or several materials and designed to transport, protect, and sell a product. For the good of the users of products contained in these packages, most countries have established certain laws and regulations that must be complied with before the goods are admitted into the particular country. This fact is accepted in theory by most U.S. exporting companies, but in practice it is too often forgotten. As a consequence, goods may be refused entry into the country of export, or unduly delayed, with a resulting loss of money and goodwill.

The reason for this state of affairs can be found in the attitude of too many U.S. companies, an attitude that says: "If our package is good for the United States, it is good also for our customer abroad." Twenty or even ten years ago, this attitude would probably have caused no difficulties. People abroad were eager to receive U.S. goods and did not care too much about how they were packaged. But today the situation has changed. With the advent of sophisticated packaging materials involving treatments, coatings, and laminations with plastics and chemicals, foreign governments—like the United States—have been forced to create specific laws and regulations to protect their consumers.

I have found that for most countries the most stringent laws and regulations are those that cover packaged foods and pharmaceuticals. And again, you cannot say: "If it is FDA, we will have no trouble in the importing country." France, for example, is very strict about some products that come in direct contact with foods—much more strict than is the United States. However, in other cases they are less difficult than the Food and Drug Administration. You therefore have to proceed with extreme caution when you ship your packaged goods, or you run the risk of being hopelessly entangled in litigation.

Marking and Labeling

Most industrialized countries have labeling regulations that apply both to the shipping container and to the package itself. Even countries like Qatar—which is certainly not a highly industrial country—have labeling regulations for imported food products.

The marking of the shipping containers in English is usually acceptable. But avoid indications that are understood in the United States but not in the foreign country. For example, the sign # can mean *number* or *pounds* here, but means nothing overseas and is a source of confusion. I recommend therefore that you check your markings with your customer or forwarding agent. As to the package itself, you should find out whether the English text is accepted. Countries like Germany, France, and Spain will accept the original labeling, provided that the measures or weights are converted into the metric system.

As a matter of fact, very often the U.S. exporter should keep the original package. For instance, a Del Monte can of sliced pineapple with a foreign language label would be considered a second-rate substitute. But if you do have to change the label in order to conform with local regulations, don't forget that your product may be re-exported and end up in another country; so here again check with your customer and make your label acceptable in both countries. You should not expect too much trouble in this matter, because many countries are aware of the language barrier and other labeling problems, and they keep their regulations down to a strict minimum.

Advertising and Promotion

If you send your packaged products abroad there is a good chance that you will want to promote them by the use of advertising. This is a good move, provided that you avoid the legal pitfalls and make sure your advertising is done within the legal framework of the importing country.

I used to have a client who wanted to sell one of his packages in Germany. Full of enthusiasm and without my knowledge, he had one of his best ads translated into German and placed in Germany's best packaging magazine. Then a month or so later he sent me the magazine with the ad. As soon as I had read it, I phoned him and told him to get in touch with his lawyer, because his ad contained a serious violation of German advertising regulations in saying: "Our product is the best in Germany." Such a statement, at least in Germany, is considered slanderous. You can say that your product is one of the best, but how can you be so sure that it is *the* best?

A week or so later, my customer received a letter from one of his competitors in Germany, saying in short: "Stop advertising that way or be sued." We promptly retired the ad with a letter of apology to the competitor, pleading ignorance—but in a court, as you know, ignorance of the law is no excuse. Therefore, I again recommend caution before you go ahead with your advertising program. Check with your client and the other sources mentioned earlier. Such a precautionary measure will not take too much time, and it may save you from many problems later on.

Trademarks and Patents

When you ship a package abroad, you should make sure that you don't infringe on any patents, designs, or trademarks in the importing country. With patents there is usually no problem, because if an identical package is patented in the importing country, either the foreign manufacturer is a licensee of the package manufacturer or one of the two patents is invalid. If the invalid patent is yours, you will have to secure a license from the foreign company or change your package.

To avoid any infringement, it is advisable to have an infringement search made in the local patent office. The results of the search should give you a good indication not only of the state of the art but also of the possible competition. With the search as a basis, if you want to be protected, a patent application for your package should be filed in the importing country; patents are good only in the country of issue, and there is no universal patent.

All countries in the free world offer the same patent protection of three basic prerogatives, preventing anyone else from making, using, or selling the invention unless the owner gives his permission or license. But in almost everything else, there are wide divergences. Terms of patents range from five to twenty years and can be dated from the filing date or from the date of issue. Fees vary from several hundred dollars to over $3,000 for the life of the patent. Most countries, not including the United States, have graduated annual payments that are low in the beginning but escalate toward the end of the patent term. Also, most countries require that the patent be "worked." Some countries require that there be no prior use or publication before filing for the patent, but a convention among most countries allows foreign filing within a year of domestic filing. Special legal assistance is advisable to get full information on the local laws and costs.

Trademarks. If you want to use your trademark abroad, you will be pleased to know that you can be fairly well protected if you exercise some caution right at the beginning. The countries of the world are broadly divided into the common law and the civil law legal systems. In the common law countries, the first to *use* a mark is the owner, and registration of the mark only gives procedural advantages to the registrant in matters of litigation in defense of the mark. In civil law countries, the first to *register* owns the mark; some U.S. companies have found their marks preempted in these countries by local businessmen who want to take advantage of the excellence and the prestige of the U.S. company, and to pass off with free advertising goods and services of lesser worth. It is therefore advisable always to register your mark as soon as you decide to go into a foreign country.

Another word of caution: when applying for a trademark in a foreign country be sure that it does not have a comical, stupid, or even obscene connotation in the language of that country. One company in Japan made use of a trademark that, translated into English, read KREEP.

Foreign Contracts

If you package for foreign markets, you will inevitably have to prepare contracts with your partner overseas. Your lawyer will prepare a contract according to your directions, but remember that an international contract must maintain a delicate balance of many factors:

Geographical. Try to know as much as possible about the country you will be doing business in. What are its size and population? What is the state of its economy? Is it an industrially advanced nation or a developing country? The more you know, the better your chances are for a successful contract.

Political. You will be dealing mostly with noncommunist nations with a democratic or not-so-democratic government. Sometimes you will find the government strongly influenced by socialist doctrines. In our definition, a government is democratic when it adheres to the five following principles:

1. Freedom and inalienable rights of man.
2. Confidence in the power of reason.
3. Limited government.
4. Preeminence of the individual rather than the state.
5. Peaceful discussion and consent to settle differences.

Many countries have what they call "nationalized sectors" in their economy, such as coal miners, steel, energy, postal services, telegraph, and telephones. You should be aware of such facts.

Cultural. It is in the cultural and psychological aspects of life that foreign people differ most from Americans. We can divide the world in four groups:

1. Anglo-Saxon and Scandinavian.
2. Germanic.
3. Latin.
4. Arabic and Oriental.

Each of these groups has a different approach to the basic aspects of life—for example, the attitude toward social mobility. Is it an "open" society like the United States or a class-conscious one? Is it striving toward educational equality? Are the family ties strong? What is the attitude toward private property and business practices? Is there a strong middle class? How deep is nationalist sentiment? A sound answer to these questions may help avoid many sometimes irreparable mistakes.

Legal. This aspect has been covered in our discussion of trademarks and patents.

A Final Word

Most foreign countries, like the United States, have packaging regulations that can affect your exported product. In some of these countries the rules are lenient; in others they are very stringent. It is not always easy to find information on these rules, but it can be done. The first people to contact are your client overseas and your forwarding agent or customs broker; however, you should probe deeper. A good source is the country's packaging institute. But too often, it is understaffed and not equipped to give you the information you need. Furthermore, you will have to conduct your correspondence in the language of the country from which you are seeking information. I would recommend that you contact the American Commercial Attaché in the country you are shipping your products to. It is his job to help you, and if he is any good, he will find the regulations that apply to your case and have them translated for you.

The fields in which packaging regulations are particularly important are those of food, chemicals, and pharmaceuticals. Most countries have their FDA equivalent, but no two countries have the same regulations. Sometimes, people here in the United States complain about the complication of the FDA and the Fair Packaging Laws. My only answer is to point to the problems of any large company in the Common Market. The exchange of goods within the Community is no longer hampered by customs, but it still has to comply with a set of five different legal structures. It will probably take a long time before a uniform law can be put into effect.

I would like to conclude by drawing your attention to three points I consider important when packaging for foreign markets: First, proceed with extreme caution and don't take anything for granted. If necessary, establish procedures and prepare a checklist that should be followed step by step. Second, use your best people in the international field and train them thoroughly. The world is your market and you will find that it can become a profitable segment of your business. Third, don't be discouraged by the complexity of the job. Robert Townsend, in his book *Up the Organization,* mentions that it is 27 times more difficult to operate in Hong Kong than in Duluth, Minnesota. But for those who want to make the effort it will prove to be very rewarding indeed.

Prospects in International Packaging Operations

David J. Freiman

THE foreign operations of American companies are becoming an increasingly important source of profit dollars. Foreign subsidiaries of American companies generated nearly $5 billion in profit after taxes in 1968 and $5.5 billion in 1969. The projections are for over $6 billion in 1970.

American industrial companies, including manufacturing, mining, and petroleum, invested $10.7 billion in properties, plants, and equipment outside the United States in 1969—an increase of 25 percent over 1968. Estimates call for an increase to $12.3 billion in 1970. Companies that manufacture products—which to packaging people is of prime interest—contributed

DAVID J. FREIMAN is manager of packaging for the Consumer Products Division of Corning Glass Works in Corning, New York.

$5 billion of the total $10.7 billion in capital spending overseas, and this figure should remain above the $5 billion mark in 1970.

It is fascinating to note that the combined gross national product of the free world's 20 leading industrial nations in 1970 will be over $2 trillion at current market prices, and the United States will have produced 47 percent of this output. But we must look with concern on the fact that the United States' contribution to *growth* in the volume of free-world output is tapering off. In 1970 the United States will contribute 20 percent to the growth of the free world's economy, ranking second after Japan, which will contribute 27 percent to growth.

Let's look, for a moment, at the world's population projections. Currently, the United States with 206 million people has 6 percent of the world's population. However, in 30 short years, the United States will have only about 4.4 percent of the world's population, or 300 million of the projected 7 billion persons who will inhabit this planet. All this adds up to the fact that by the year 2000, 6.7 billion persons will be living in foreign markets and, we hope, enjoying the benefits of a healthy economy. As a result, more and more companies are becoming multinational, and they must be prepared to capitalize on the growing opportunity in foreign markets. They can do this in two ways: (1) by gearing their exports to meet the product and packaging needs of the foreign markets, or (2) by increasing the activities of the foreign subsidiaries to serve those needs. It is to the second alternative that I shall address myself here.

It would seem that the long-term benefits are greatest in the development of foreign subsidiaries to serve the needs of their own markets. If this is the case, it is the role of the parent company to develop the foreign subsidiaries' expertise in manufacturing, product development, merchandising, and marketing, so that they can operate efficiently and generate maximum profits.

Foreign subsidiaries are generally of two types: new enterprises and existing operating units that have been acquired by an American company. In a newly created enterprise owned by a U.S. company, the upper echelon of management is usually trained in the parent company's methods of operation, and the organization of functions such as packaging—as far as structure, training, and packaging strategy are concerned—is usually done before operations are started out of the country.

When the U.S. product and the foreign product are similar, it is possible to develop packaging for a foreign subsidiary that is as good as or better than that used by the parent company, right from the initial introduction into the market. Until foreign packaging personnel can be trained, it is necessary to have an American packaging manager who knows the foreign market, knows the product, knows the sources of supply, and has the respect and confidence of top management of the foreign subsidiary. The engineering and design can be obtained locally, but it will take a packaging professional to find the right basic packaging concept for the product and then to supervise the execution of the concept properly. This is a skill that is rare in the United States and with few exceptions almost nonexistent outside the United States.

The role of the American packaging manager in a situation like this is almost the same as the role that he has in the parent company. He is responsible for the initial packaging used for his foreign subsidiary, although, as the initial packaging is developed, personnel of the foreign company can be trained to assume this developmental responsibility. From the corporate standpoint, utilization of this technique for package development for foreign subsidiaries provides immediate, maximum efficiency with regard to both cost and package effectiveness. The following case study of our European subsidiary, Corning Nederlandse Fabrieken, is a good illustration of a major foreign package development completely directed by the American package manager of the parent company involved.

Establishing the Foreign Package

In 1965, Corning Glass Works formed a subsidiary company headquartered in The Netherlands, called Corning Nederlandse Fabrieken. This company was formed to manufacture, market, and merchandise Corning Ware® cookware on the continent. The only differences between the U.S. product and the product to be produced by C.N.F. were a slight variation in product decoration and a substitution of the name "Pyroflam" for "Corning Ware." We decided on this change in the product name because we felt there was no equity or recognition of the name Corning in Europe. The name Pyroflam seemed to be more descriptive of the product benefits. The marketing strategy indicated that the package used for Pyroflam would be extremely important, serving as a major merchandising tool at the point of sale to establish the product line identity with the consumer.

The first stage in package development is to establish the package objectives. The objectives we established for the Pyroflam package were as follows:

1. Sufficient protection to enable the product to be moved from its origin of manufacture through the distribution system and to get to the consumer intact.
2. Surface graphics that would give strong product identity.
3. Graphics that would suggest good connotations of cooking to the consumer.
4. Multilingual graphics that would give a gift connotation.
5. An economic package that would lend itself to automatic forming.

It should be noted that these general objectives are similar to our U.S. Corning Ware packaging objectives and are international in character.

For a short time, our French subsidiary had unsuccessfully attempted to market Pyroflam in a folding carton. When we evaluated this package against the list of objectives, we found that its major deficiencies lay in a lack of surface graphics that would give strong product identity. In the United States we had found that the use of color was a good way to get this identity; the blue used for Corning Ware is referred to by many consumers as "Corning Ware blue." Our first inclination was to use a modification of our U.S. package, but after conducting a retail survey, I found that there were many blue packages used for housewares. We would get no exclusive identity through the use of this color.

The package finally developed for Pyroflam utilized a distinctive pink color with the product logotype in a ring of flame. Copy on the package was restricted to the catalog number on all panels of the package to aid in stockroom handling, the capacity of the product in centimeters, and a silhouette of the product. By restricting the copy, we had a package that was multilingual and international in character, because it could not immediately be identified with its country of manufacture. This package, which had achieved point-of-sale product identity through the use of the Pyroflam pink color, proved extremely successful and is still being used. Consumer research concluded that respondents identified the Pyroflam package with some form of cookware, even when they did not know what the specific product was. This is an important package benefit when a product is being introduced into a new market.

Because of the mechanical strength characteristics of Pyroflam, the package for this product utilized corrugated board and an extremely economical machine-formed carton construction. All packaging components were successfully purchased in The Netherlands and met all purchase specifications. Now that the packaging format was established, it became a simple procedure to adapt it to new items as they were added to the line, and our personnel at C.N.F. were able to carry out this responsibility.

My own assignment in developing the packaging program was complete once the packaging format was developed and approved, and my role has reverted to acting as a consultant as the need arises. If I have done my job well, if I have developed a good line of packaging that is versatile enough to be long enduring, and if I have properly trained our people, I will need to spend very little time assisting this subsidiary.

Handling Problems in the Foreign Subsidiary

When dealing with foreign subsidiaries that have been acquired and have previously existed as an operating unit, often one has to contend with problems that have been built up to a critical level through long-term nonmanagement of the packaging function. In such situations, it takes a long time to reverse the effects of poor packaging and reduce the monies wasted through packaging inefficiencies. The application of package management techniques such as value analysis is relatively new in the United States and almost unknown overseas.

We can usually assume that the American packaging manager is not called in to consult with the foreign subsidiary until problems have come to the attention of top management at both the parent company and the foreign subsidiary. The American packaging manager is usually in a sensitive position. He must come in as a foreigner, often as a stranger who cannot speak the language, and develop a working relationship with the personnel of the subsidiary. This requires creating an atmosphere of trust, mutual respect, and understanding between the foreign subsidiary people and the American packaging manager—even though he is criticizing much of what has been done in the past. To accomplish his goal, which is to improve the packaging of the subsidiary, the American packaging manager must be not only an expert on packaging but also a diplomat.

At best, all the consultant can expect to accomplish is to outline packaging needs, suggest solutions to current problems, and show how these solutions can be implemented. The actual job of implementation must be the responsibility of the foreign subsidiary personnel. Unfortunately, trained packaging personnel are very often unavailable to follow through and execute recommendations optimally. A large part of the American packaging manager's role when dealing with foreign subsidiaries is to train their personnel to become competent in techniques of package development. This is one reason for the long time that it takes to reverse the effect of long-term nonmanagement of the packaging function.

However, in spite of the negative aspects of the situation, major progress can be made quickly; a Parkinsonian type of law is in effect that states, "The degree of progress is in direct proportion to the amount of chaos present." A little well-directed packaging effort goes a long way in gaining benefits.

When consulting with our foreign subsidiaries, these are my ten commandments of how to work with them:

1. Gain a commitment on the part of the foreign subsidiary by charging to it all expenses incurred by the visit. (This includes the option to charge the consultant's salary, although I have never seen fit to do this.)
2. Set down in writing, at the beginning, the specific objectives of the consultations, and reach an agreement with the principals involved on the validity of these objectives.
3. Establish a work schedule relative to the time allotted to the project—a schedule that will be adhered to even if it means working nights and weekends. Because of responsibilities with the parent facility, overseas trips should be limited.
4. Thoroughly study the situation before making any comments or suggestions.
5. Make no criticism without giving possible solutions to problems and ways of implementing them.
6. Give a verbal report to the principals involved before returning to the United States, and try to satisfy the objectives stated at the beginning of the project.

7. Write a formal report covering in detail all aspects of the study, including an investigation of possible alternatives that you have considered and worked on after your return to the States.
8. Get confirmation in writing that the subsidiary accepts the recommendations as covered in the report.
9. Check the subsidiary's progress in following up on the recommendations.
10. Assist the subsidiary with services, information, or guidance to enable it to carry out the recommendations.

After the initial consultation, make it a policy to continue consultation only with foreign subsidiaries that accept and implement the recommendations.

Index

Index